Explor

FireWorks

Series editors:

Gargi Bhattacharyya, Professor of Sociology,
University of East London

Anitra Nelson, Associate Professor, Honorary Principal
Fellow, Melbourne Sustainable Society Institute,
University of Melbourne

Wilf Sullivan, Race Equality Office, Trade Union Congress

Also available

Reinventing the Welfare State:
Digital Platforms and Public Policies
Ursula Huws

Pandemic Solidarity:
Mutual Aid During the Covid-19 Crisis
Edited by Marina Sitrin, Colectiva Sembrar

Exploring Degrowth
A Critical Guide

Vincent Liegey and Anitra Nelson

Foreword by Jason Hickel

PLUTO PRESS

First published 2020 by Pluto Press
345 Archway Road, London N6 5AA

www.plutobooks.com

Copyright © Vincent Liegey and Anitra Nelson 2020

Every effort has been made to trace copyright holders and to obtain their permission for the use of copyright material in this book. The publisher apologises for any errors or omissions in this respect and would be grateful if notified of any corrections that should be incorporated in future reprints or editions.

The right of Vincent Liegey and Anitra Nelson to be identified as the authors of this work has been asserted in accordance with the Copyright, Designs and Patents Act 1988.

British Library Cataloguing in Publication Data
A catalogue record for this book is available from the British Library

ISBN	978 0 7453 4201 6	Hardback
ISBN	978 0 7453 4202 3	Paperback
ISBN	978 1 7868 0759 5	PDF eBook
ISBN	978 1 7868 0761 8	Kindle eBook
ISBN	978 1 7868 0760 1	EPUB eBook

Typeset by Stanford DTP Services, Northampton, England

Contents

List of Figures and Boxes — vi
List of Abbreviations and Symbols — vii
Glossary — ix
Series Preface — xiv
Foreword by Jason Hickel — xv
Preface — xix
Acknowledgements — xxiv

1. Introduction: Exploring 'Degrowth' — 1
2. Decolonising Our Growth Imaginaries — 20
3. Degrowth in Practice — 49
4. Political Strategies for Degrowth — 86
5. The Degrowth Project: A Work in Progress — 116

Postface: 'Now Is the Time of Monsters' — 154
Appendix 1: A Platform for Degrowth — 158
Appendix 2: The Content of the Unconditional Autonomy Allowance — 163
Appendix 3: Implementing the Unconditional Autonomy Allowance: Transitionary Steps — 168
Notes — 170
Selected Further Reading and Links — 186
Index — 189

List of Figures and Boxes

FIGURES

1.1	Two imaginaries	13
2.1	Energy descent	29
2.2	Brands aren't your friends	38
3.1	Seven types of ordinary happiness	61
3.2	Fifth International Degrowth Conference banner	75
4.1	Activist Patrick Jones at a climate change rally, Melbourne 2019	93
4.2	Advertising	108
4.3	Pete Seeger – 'If it can't be reduced…'	109
5.1	Fuck work	118
5.2	AdHack Manifesto	135

BOXES

2.1	The tourist and the fisherman	40
5.1	A lifestyle menu	144
5.2	Degrowth formation: Cargonomia	146

List of Abbreviations and Symbols

aka	also known as
CADTM	Committee for the Abolition of Illegitimate Debt
DIO	do it ourselves
DIY	do it yourself
EU	European Union
FCAI	Federal Chamber of Automotive Industries
GDP	gross domestic product
GPI	genuine progress indicator
IPBES	Intergovernmental Science-Policy Platform on Biodiversity and Ecosystem Services
IEA	International Energy Agency
kWh	kilowatts per hour
L	litres
LETS	labour/local exchange trading system/s
m	metre/s
MAI (in French, RMA)	maximum acceptable income (*revenu maximum autorisé*)
NIMBY	not in my back yard
PPLD	Collectif Nous Sommes Parti.e.s Pour La Décroissance
R&D	Research and Degrowth

UAA	unconditional autonomy allowance
UFO	unidentified flying object
UK	United Kingdom
US	United States (of America)
ZAD	Zone à Défendre

Glossary

Autonomy A central political concept for the degrowth movement, mainly influenced by Cornelius Castoriadis' work, where agency and subsidiarity are central. Refers to the will, capacity and capability to self-define, to decide and act responsibly within socio-political limits, whether as an individual or a collective. Its opposite, heteronomy, is effective submission, or giving power over, to an external authority.

Commons and *commoning* Refer to co-created, co-governed and co-accessed cultural and natural resources and associated activities. The key questions for degrowth, in particular for the degrowth project, focus on creating and experimenting with new types of co-governance which are significantly more equal, transparent, democratic and sustainable than those driven by the logic of the market.

Convivial A central concept for degrowth, coined and developed by Ivan Illich. As a practice, it means a cooperative, mutual, sociable and sharing approach. As a characteristic of societies, it means that citizens (not experts or technocrats) directly control technologies and institutions, so that technologies are selected to serve the common interest and do not breach the counter-productivity threshold.

Counter-productivity threshold The point at which the human time and effort, and material and energy costs, of a technology surpass the saving made by its use in a comparison with similar techniques or technologies used to achieve the same result.

Decolonisation of the imaginary A central concept and objective of degrowth, first, as a provocative slogan and, second, as a multidimensional matrix of thoughts and theories. Refers to processes involving the deconstruction of concepts and beliefs associated with growthism in order to liberate and open practical pathways to sustainable and desirable lifestyles and models of society.

Ecofeminism Ecological feminist perspectives exploring and advocating emancipation from Western patriarchal domination of both women and ecosystems. A key principle of degrowth applied by acknowledging needs for gender equality, engagement of all in care of people and nature, thus deconstructing gender identities and roles associated with domination and exploitation.

Ecological economics A transdisciplinary field of studies addressing contemporary challenges by focusing on governance of economic activities to achieve human well-being, ecological sustainability and social justice (recently aka 'social ecological economics').

Frugal abundance The only abundance for the degrowth movement is a frugal one. Western productivist societies, driving to overcome limits, are characterised by 'overabundance'. Philosophically abundance, like well-being, is subjective and, as a desire, culturally transferable. The degrowth movement advocates for practices, philosophies and livelihoods that respect limits but are rich and full in meaning. Also referred to as 'happy sobriety'.

Glocal The application of universal (global) principles locally.

Low-tech approach A provocation to rethink human relationships with technologies – as in Ivan Illich's 'tools for conviviality' and Jacques Ellul's 'technician totalitarianism' – this means inventing, developing and sharing the most simple, local, easy-to-implement and easy-to-use tools or technologies to satisfy our basic needs. Includes questions around how to design tools in sustainable, open-source, shareable and autonomous ways.

Municipalism A movement advocating direct democracy and subsidiarity, power at the local (municipal) level, often at the expense of – or in a confederated form as a substitute for – the state. Murray Bookchin advocated a libertarian form. Municipalism is practised by Kurdish communities in the liberated territory of Rojava (northern Syria) and the highland Zapatista communities of Chiapas (Mexico). Influential in many contemporary movements, such as commoning and degrowth.

Open relocalisation Localisation of production and distribution deemed necessary for environmental, energy-related and social reasons. Means re-appropriating knowledge and skills for low-tech and ethical local decision-making, that is, direct democracy. In contrast to reactionary protectionism open relocalisation calls for reversing globalisation in transparent, all-inclusive ways based on solidarity with the aim of diversifying communities and environments to enrich life.

Planned obsolescence A common and wasteful characteristic of capitalist production whereby new models of standard products embody features that make older versions less useful and/or less attractive, so that although they may still be functional and useful, 'outdated' models are discarded.

Pluriverse and *pluriversity* Refer to saving, respecting and re-creating diversity in a globalised world without reducing solidarity and exchange but rather offering a way for degrowth to find the right balance between universal rights and risks of Western cultural colonialism.

Post-development Central to degrowth, a strong critique of colonial, culturally imperialistic beliefs associated with the Western model of 'civilisation' imposing 'development' globally with devastating cultural and environment impacts. Constructively, an invitation to acknowledge, learn from and engage with the wisdom, skills and know-how of non-Western cultures, languages and civilisations.

Re-embed the economy Karl Polanyi's approach, highlighted by Serge Latouche, for reversing the socio-economic and cultural centrality of the market, the economism, in productivist societies. Central to the degrowth movement's quest to decolonise the dominant economically 'rational' imaginary, to achieve post-capitalism via decommodification and economic regulations.

Transhumanism A philosophical movement promoting the development and broad accessibility of sophisticated technological enhancements for human minds and bodies.

Series Preface

Addressing urgent questions about how to make a just and sustainable world, the Fireworks series throws a new light on contemporary movements, crises and challenges. Each book is written to extend the popular imagination and unmake dominant framings of key issues.

Launched in 2020, the series offers guides to matters of social equity, justice and environmental sustainability. FireWorks books provide short, accessible and authoritative commentaries that illuminate underground political currents or marginalised voices, and highlight political thought and writing that exists substantially in languages other than English. Their authors seek to ignite key debates for twenty-first-century politics, economics and society.

FireWorks books do not assume specialist knowledge, but offer up-to-date and well-researched overviews for a wide range of politically-aware readers. They provide an opportunity to go deeper into a subject than is possible in current news and online media, but are still short enough to be read in a few hours.

In these fast-changing times, these books provide snappy and thought-provoking interventions on complex political issues. As times get dark, FireWorks offer a flash of light to reveal the broader social landscape and economic structures that form our political moment.

Foreword

This book arrives at just the right time. A few years ago, degrowth was a relatively fringe idea. You might have encountered it among ecological economists and post-development thinkers, and it was being picked up here and there in certain activist communities. But suddenly it's bursting into the mainstream in a way that nobody anticipated. People are eager to learn what it's all about.

Part of this has to do with growing public awareness of the ecological crisis. It's clear to any rational observer that something is terribly wrong. Rainforests are going up in flames, heatwaves are searing across the poles, ice sheets are melting faster than anybody imagined; as the climate crisis worsens, and as our existing attempts to halt it fail, people are scrambling for new ideas.

We know what we need to do. The Intergovernmental Panel on Climate Change (IPCC) insists that in order to avert dangerous climate breakdown we need to cut global emissions in half by 2030 and reach zero by 2050. And, of course, high-income nations need to move much more quickly than this, given their outsized responsibility for historical emissions.

It would be difficult to overstate how dramatic this trajectory is. We've built up a global fossil-fuel infrastructure over the past 250 years, and now we have to completely overhaul it in only 30. Everything has to change, and fast.

Extinction Rebellion, Fridays for Future, the Sunrise Movement – people everywhere are now rallying to pressure politicians to act. The old fantasy that market mechanisms will somehow magically solve the climate crisis has been thoroughly dashed, and a new consensus is emerging: we need coordinated government action on a massive scale to accomplish a rapid transition to renewable energy.

But there's a problem. Climate scientists are warning that it's not possible to transition fast enough to stay within the carbon budget for 1.5°C or 2°C if the global economy continues to grow at the usual rates. Why? Because more growth means more energy demand, and more energy demand makes it all the more difficult to roll out enough renewable capacity to meet it.

This is why we have failed to make any progress against climate breakdown so far. There have been some extraordinary gains in renewable energy production over the past two decades, but all that new energy is being added *on top of* fossil fuels, rather than replacing them. Why? Because growth is outpacing our ability to roll out solar panels and wind turbines. We are fighting a battle we're doomed to lose.

For years, policymakers and economists have papered over this problem by promising that technology will come to the rescue: we can carry on with growth as usual, overshooting the carbon budget, because negative emissions technologies will suck emissions out of the atmosphere later in the century. But scientists have rejected this idea as a farce: to deploy negative emissions at scale is not only unfeasible, but tremendously risky: if it fails, we'll be locked into a high-temperature pathway from which we will never be able to escape.

It's time for a different approach. If we want a decent shot at climate stability, high-income nations will need to actively scale down unnecessary economic activity. Less growth means less energy demand, making it easier for us to accomplish a rapid transition to renewables, while taking pressure off our living planet in the process.

There's a striking consensus forming in this direction. In 2018, 238 scientists called on the European Commission to abandon growth as an objective and explore post-growth futures. In 2019, more than 11,000 scientists from over 150 countries published an article in the journal *BioScience* with much the same message. As for the IPCC, its lead scenario for staying under 1.5°C calls for a significant reduction in energy and material throughput – the mainstay of degrowth thinking.

This might seem scary, on the face of it. Most of us assume that growth is essential for human well-being. But, remarkably, there is no empirical evidence for this claim. Beyond a certain point, which high-income countries have long since surpassed, the relationship between GDP and welfare completely breaks down. In fact, evidence is now piling up to show that the dogged pursuit of growth is beginning to cause social *bads*.

The key thing is to recognise that we already live in an abundant economy. The problem isn't that there's not enough for everyone to flourish, the problem is that it's all captured at the top. We can improve people's lives right now, without any growth at all, simply by sharing what we already have more fairly and investing in public goods. Equity is the antidote to the growth imperative.

Of course, all of this leaves us with a million questions. What does a such an economy look like? How does it work? How do we get there? The book you hold in your hands traces a path through this thicket, and paints a picture of the new economy that lies ahead – an economy that enables human flourishing for all within planetary boundaries.

We stand at an inflection point in world history. We know that a better world is not only possible; it is waiting to be born.

Jason Hickel
10 March 2020

Preface

While writing this manuscript, we experienced – with you – the warmest January ever recorded for planet Earth. The planet's ten warmest Januaries all happened after 2002. According to the US National Centers for Environmental Information: 'January 2020 marked the 44th consecutive January and the 421st consecutive month with temperatures, at least nominally, above the twentieth-century average.' Certain European countries were particularly badly hit; Northern Hemisphere January temperatures made it the warmest on record at 1.5°C above average.[1]

Vincent, in Europe, sweltered through the hottest of summers in 2019. Between 1 May and 30 August 2019 almost 400 temperature records were broken across 29 European countries. The Berkeley Earth climate institute (California, US) detected 1,200 locations in the Northern Hemisphere where residents experienced the hottest ever conditions in the given month. An all-time-high temperature of 46°C in France was accompanied by new highs in the Netherlands, the UK, Belgium, Germany and Luxembourg.[2] Then, seemingly perversely, the winter of storms hit. Following quick on the heels of Storm Ciara (8–9 February 2020) was Storm Dennis (15–16 February 2020), causing heavy rain, strong winds and flooding that impacted most of Britain – prompting 'a record-breaking number of Environment Agency flood warnings and alerts in England on Sunday'.[3]

Meanwhile, in the Southern Hemisphere, Anitra experienced the historic bushfires now known as the Black Summer. To backtrack, Australia is renowned for bushfires. The 2009 Black Saturday fires in the state of Victoria are authoritatively estimated to have 'released energy equivalent to fifteen hundred times that of the atomic bombs dropped on Hiroshima'. Australian environmental historian Tom Griffiths continues:

> Of the 173 people killed on Black Saturday, two-thirds died in their own homes. Of those, a quarter died sheltering in the bath. There were relatively few injuries: the annihilation was total, and the day after brought an awful stillness and silence. The wind change was a killer, but if it had not arrived when it did the Kilmore East fire might have swept into the thickly vegetated suburbs of Melbourne's north-east.[4]

Melbourne is the capital of Victoria and in 2015 the Australian Bureau of Statistics counted more than half a million residents living in those suburbs. Around 90 per cent of Black Saturday fatalities occurred in fires started by faulty electric powerlines associated with privatisation of the state's electricity supply.

But there was worse to come. Fast forward and Black Summer 2019–20 was exceptional. Between July 2019 and February 2020, mainly due to a long period of drought and record temperatures, close on 16 million hectares of land and vegetation were burnt, some in mega-fires creating their own weather systems. A lightning strike late in October north west of the capital city of Sydney started the largest recorded bushfire in Australia. By mid-December, a month in which temperatures were persistently around 3°C above the average, it joined and absorbed independently started fires. The social and ecological impacts are hard to capture and are still playing out.

A few estimates indicate the destruction and likely repercussions. Animal deaths were over 1 billion, insect deaths multiple billions. Of the 33 deaths due to the fire, several were firefighters. Several small communities were all but decimated as 3,500 homes were lost and thousands more partially damaged.[5] The smoke and air pollution swathed the continent and moved across the globe. As the fires raged through January, the federal government made its first ever compulsory call out for Reservists, engaging 6,500 Australian Defence Force personnel to support response efforts, which led to a Royal Commission into National Natural Disaster Arrangements.[6] We live in extraordinary times. More significantly, we are in times where that 'extraordinariness' is expected to become, to be, normal. Indeed, the fires were followed by record-breaking rain and floods. Mercifully, that water put out fires that were still burning but the floods brought further destruction.[7]

As we wrote passages on the pedagogy of such catastrophes, we noticed how such unnatural disasters of the Capitalocene were heightening people's awareness, opening them to new understandings and debates, especially questioning the causes. It is clear that many had expected that climate change and global warming might only affect the disadvantaged, or might only affect themselves in the distant future, or even just impact on future generations. It seems that many had sidelined worrying about challenges that they assumed others, such as governments or industry, would solve.

But, now, with water, winds and flames destroying lives and homes, many feel immediately vulnerable and a sense of urgency has enveloped electorates and communities across the planet. Declaring a climate emergency has been one response from local councils to national governments. But there is a strong appetite, too, for appropriately responding to the multidimensional impacts of growthism. People understand that climate and other environmental changes are happening to us, and, if we don't manage the transformations necessary, we will be victims.

As if to make that point, as we finalised this manuscript, Europe's storms and Australia's fires became 'yesterday's news'. In these times of multi-various accelerations in change, we cannot predict anything. Suddenly, the Coronavirus, which started in China, has not only spread right around the world and resulted in more than 2 million cases and around 130,000 deaths by 16 April 2020 – but also the calamity slowed our 'just-in-time' global economy. Capitalism is like a spinning top: as it slows, it falls over, its life and function dead. Economists forecast a global crash and drawn-out global depression of an unforeseen magnitude. Governments have moved to shore up the system with trillions of dollars of funds as under-resourced and impoverished national health sectors are overwhelmed by the pandemic.

The world hangs on the announcement of a vaccine or treatment that will turn all this into an easily forgotten nightmare. Similarly, the elites still crave the latest 'whizz-bang' technological solutions to cutting carbon emissions and achieving a zero-carbon economy. All such perspectives fail to recognise that climate change, and global heating due to the rise in carbon emissions, are but symptoms and not the cause. In terms of Earth's regenerative capacity, we are over-producing and over-consuming 'as if there were no tomorrow'.

As the planet's ecosystems erode and politicians obsess over growth objectives, what we really need is degrowth, that is, to transform our ways of living, our economies, our polities and our cultures to live within Earth's limits. This is what our book is about – the degrowth movement.

Vincent Liegey and Anitra Nelson
16 April 2020

Acknowledgements

Pluto Press encouraged us to write this work for the FireWorks series, after the editors of the series identified topics that demanded summary explanations for interested public and academic readers. Thus, we thank series editors Gargi Bhattacharyya, Wilf Sullivan and editorial director David Castle for supporting the development of the book. We also thank the four anonymous peer-reviewers of our initial proposal to Pluto Press, who encouraged us by recommending publication and by pointing out ways we could improve the proposed work. Finally, among those who contributed to the work's development, we especially thank Jason Hickel for his Foreword, and Serge Latouche for his supportive words.

As a movement that evolved in Europe, in particular France, 'degrowth' initially developed a literature and circulation in non-English languages. In order to introduce this highly relevant contemporary concept and movement to English-speakers and inform them more fully, Welsh-Australian activist-scholar Anitra Nelson worked with prominent French degrowth advocate, activist and independent scholar Vincent Liegey. Our main acknowledgement is to Vincent Liegey's previous co-authors Stéphane Madelaine, Christophe Ondet and Anne-Isabelle Veillot – and their publisher Les Éditions Utopia (Paris, France) – for allowing us to draw on *Un Projet de Décroissance: Manifeste pour une Dotation Inconditionelle d'Autonomie* (Degrowth Project: Manifesto for an Unconditional Autonomy Allowance), published in 2013 (see the website A Degrowth Project, listed in Further Reading).

We gratefully acknowledge the contributions of various readers of our first draft manuscript who brought a series of errors and gaps to our attention. Many thanks to Silvio Cristiano, Mladen Domazet, Dan Golembeski, Terry Leahy, Alex Panelli, Bronwyn Silver and Lee Tan. As is customary, we take full responsibility for all the weaknesses in this manuscript.

We gratefully acknowledge permission received from Australian cartoonist Michael Leunig for the cartoons appearing as Figure 1.1 Two imaginaries and Figure 3.1 Seven types of ordinary happiness. Discover work by Michael Leunig here: www.leunig.com.au/

We gratefully acknowledge permission received from artist-activist Patrick Jones (Australia) for his graphic reproduced as Figure 2.1 Energy descent. Find out more about Patrick here: http://theartistasfamily.blogspot.com/

We gratefully acknowledge the graphic art of the anti-advertising Special Patrol Group (UK) reproduced as Figure 2.2 Brands aren't your friends and Figure 5.2 AdHack Manifesto. For more, see: @specialpatrols (Facebook) and @SpecialPatrols (Twitter).

We gratefully acknowledge graphic artist Eszter Baranyai and the 2016 Budapest Degrowth Conference Support Group for permission to publish Figure 3.2 Fifth International Degrowth Conference banner.

We gratefully acknowledge photographer, Meg Ulman (Australia), and subject Patrick Jones for permission to reproduce the photograph appearing as Figure 4.1 Activist Patrick Jones at a climate change rally, Melbourne 2019.

We gratefully acknowledge artist Darren Cullen for reproducing his graphic as Figure 4.2 Advertising. Find more work by Darren Cullen (UK) here: www.spellingmistakescostlives.com/

We gratefully acknowledge the graphic rendition of lines from Pete Seeger's song 'If it can't be reduced' in Figure 4.3, an image widely shared on twitter, here from a Tweet 20 May 2019 from @ViridorUK.

We gratefully acknowledge artist Josh MacPhee (US) for reproducing his graphic as Figure 5.1 Fuck work. Find more work by Josh MacPhee here: https://justseeds.org/artist/joshmacphee

We gratefully acknowledge the inclusion in chapter 5, Box 5.1 of an extract on a Swiss one planet lifestyle menu from a 20-page pamphlet *A Proposal* issued by New Alliance (see their website – https://newalliance.earth/a_proposal.pdf).

Last, but certainly not least, we gratefully acknowledge the contributions of all the degrowth pioneers and friends who have participated with us in this wonderful journey of creativity, conviviality and hope towards sustainability and enjoyment of life, in particular the collective Nous Sommes Parti.e.s Pour La Décroissance in France, the Cargonomia team in Budapest, the Support Group of the International Degrowth Conference team and, of course, our families and friends, such as Orsolya, Françoise, Daniel, Géraldine …

CHAPTER ONE

Introduction: Exploring 'Degrowth'

Protests erupted around the world in the 1960s and 1970s to highlight international civil rights, anti-war, feminist, gay liberation and student concerns along with a range of environmental and anti-consumerist issues. Protesters yelled 'NO!' – no to sending young soldiers to Vietnam, no to nuclear weapons, no to lower wages for women, no to laws against homosexuality, no to developments destroying pristine nature, no to chemical pollution of air and waters, no to universities closed to the disadvantaged. Unemployment rose as young people rejected work in dangerous and anti-social industries and institutions. An 'underground', anti-systemic movement attracted them either to the countryside – to establish alternative forms of self-provisioning – or to squat in the cities.

Simultaneously a culture of revolt became rife. Urban streets were riddled with graffiti and posters. Theatres were enlivened with spectacular, seditious and unconventional performances flouting post-war norms. Bookshops and cinemas became sources of 'banned' materials until censorship weakened and gave in. Journalism and writing evolved novel forms of creative non-fiction, discontinuous narrative and performance poetry. The young stepped out in direct actions hailing new forms of citizenship and relationships. Non-hierarchical organising and networking evolved new politicking that endured and morphed with new media technologies.

As all kinds of movements proliferated, changes in laws, policies and everyday culture ensued. Consequently, progress was made on many socio-political and cultural fronts, yet the world's ecological challenges and social inequities have deepened and expanded. Climate change is just the tip of the environmental-crises iceberg. The first couple of decades of the twentieth century have brought severe biodiversity loss and planetary apocalypse to everyone's lips. These existential challenges have been met by competing solutions such as green and circular economies, ecosocialism, other sustainability 'fixes' and universal sustainable development goals.

It is in this context of heightened debate and widespread dismay that the degrowth movement sprang to life in Europe and spread further afield. The term '*décroissance*', later translated into 'degrowth' in English, began as a provocative slogan used by activists in the early 2000s. The French political scientist and editor Paul Ariès has referred to degrowth as a 'missile word', intentionally making people question the 'growth is good and more growth better' flag under which all nations seemed to have united in economic terms.[1]

In strict translations of '*décroissance*', going beyond growth means reducing or decreasing. Proponents focus on reducing environmental use and abuse, yet degrowth is, at once, both a qualitative and a quantitative concept. The qualitative dimension is captured in concepts such as 'frugal abundance', which connects 'conviviality' – enjoying one another's company and acting in solidarity

– with valuing the richness of simplicity as in 'small is beautiful'.[2] Beyond significant misunderstandings arising externally, degrowth has developed multiple meanings and nuances within the activist movement campaigning for it.

Most significantly, the word 'degrowth' has misled to the extent that its prefix and association with words such as *de*cline and *di*minish seem to indicate that *de*growth means austerity, puritanism and even poverty. The minimalist simple-living aspect of degrowth seems to confirm such suspicions. Especially since the global financial crisis broke during 2007–8, with persisting consequences, degrowth sounds unsettling. In contrast, degrowth theorists and activists see degrowth as establishing secure and safe lives, fulfilling everyone's needs in collaborative and collective ways, as celebratory and convivial.

The degrowth principle of living within Earth's regenerative limits in socially equitable and collectively supportive ways addresses both global and environmental crises. This book is intended as an introduction to degrowth for anyone unfamiliar with the movement. Equally, it is written for those who are familiar with degrowth but would like a handy *résumé* on what the movement stands for, what it has achieved and where it might go in the 2020s. It will explain the intended meanings of degrowth for its protagonists and advocates, who have realised certain degrowth ideas and principles in mini-experiments with collective living, working collaboratively and self-governing using consensual decision-making. Chapters focus on various aspects of degrowth in action. Activists are mobilised by theories and visions, and propose policies for immediate implementation as well as establishing degrowth in stages and holistically, that is, a degrowth project.

CHALLENGING GROWTH

Challenging economic growth as a concept or ideal is neither novel nor extraordinary. However, recent critics of growth, such as the late democratic socialist Erik Olin Wright, tend to counsel market-based reforms rather than a revolutionary response aimed at minimising 'stuff' produced. Many critics of growth adopt a preference for the term 'development'. Consequently, development has become 'a word for all seasons', meaning whatever growth critics want it to mean in the circumstances, with different types of emphases on qualitatively improving the conditions of living for the majority who currently live more precariously and powerlessly than the elite few with wealth and political influence.

A reformist Western concept and practice, the whole idea of development has been rebutted in ways aligned with degrowth thinking, focusing on decolonisation and liberating imaginaries, since the 1980s in a radical 'post-development' critique.[3] Meanwhile, capitalism has grown extensively and intensively, invading new territories, new sectors and creating its very own context for growth in patents and copyrights for novel technologies, in short an information 'territory' within which to expand assets. So it is clear that calls for keeping capitalism on the more qualitative tracks of development consistently failed.

The nineteenth-century revolutionary Karl Marx's *Capital* (vol. I, 1887 [German, 1867]) has been the outstanding reference for theorists pondering the anomalies of a politico-economic system forever spiralling upwards in monetary terms, imperially expansive in its impulses and aimed at making profits for the few. Yet those who established communist regimes in the twentieth century, ostensibly to change the world according to Marx – who would have been horrified at the results – just seemed to produce another version of inequity and, significantly, economies based on productivist notions of growth.

Since the global financial crisis the most humane journalists and left-minded politicians have tended to focus on managing growth following development principles of more just distribution, at least in the 'good times'. Still, when economies turn bad, the state has been just as likely to resort to seemingly necessary austerity. In stark contrast, on the streets, in underground cultures and oppositional media, anti-capitalist demands to occupy (potentially everything) and calls for 'system change not climate change' have become rampant. In this context, it is no surprise that an explicitly anti-growth, indeed de-growth, movement would gain attraction.

In terms of the flagrant abuse of planet Earth, we know that capitalist production and trade has increasingly outstripped its regenerative capacity for the last 50 years. By 2019 this meant exploiting natural resources as if there were 1.7 Earths.[4] Much over-consumption has occurred in the Global North – where the degrowth movement started and maintains its greatest support – in Europe. Environmental crises are inextricably linked to economies harnessed to growth. Initial responses to degrowth and debates around the concept tend to confirm the extent to which our minds, our imaginaries (not simply our everyday practices) have been colonised by the idea of growth. It is as if economies without growth are impossible to imagine. Even to mention degrowth in mainstream everyday situations seems idiotic and illogical, at least until one learns its nuances, foundations and intents.

By way of one significant example, unions are structurally oriented to increasing the size of the capitalist pie, not only their slice of it. Mainstream workers and unionists most strongly identify as a class apart and opposed to capitalists and managers with their primary goal as a fairer distribution of output. Even if unions have gone on strike in environmental protests, the holistic idea of degrowth challenges their everyday struggles to maintain full employment and to gain higher wages and salaries. Indeed, the degrowth movement evolved to expose this entrenched omnipotence of the concept, practice and quasi-theology of growth. As Kenneth Boulding said: 'anyone who believes in indefinite growth in anything physical, on a physically finite planet, is either mad – or an economist'.[5]

Yet progressive and forward-thinking unions have strong campaign synergies with degrowth when they institute 'just transition' programmes for workers, move into developing arrangements for sharing work, prioritise improving the terms and conditions for part-time workers and have long-range plans for progressively shortening the average working week. Even as degrowth hits a brick wall with conventional structures and institutions, chapter 3 will show how many values and visions degrowth shares with various twenty-first-century movements such as ecofeminism, Occupy and municipalism, and with associated principles, such as autonomy, conviviality and frugal abundance.

Moreover, the rest of the book shows how degrowth action and theories have developed, cultural distinctions in degrowth's evolution in various spaces, and political controversies at the heart of the movement. As an entrée, this chapter gives an introductory tour, starting from the French political and intellectual debates that founded degrowth through to its translation into other languages, including English, that duplicated misunderstandings and led to subtle reinterpretations. How has this missile word been used? What are its drivers and its limits? And, why do debates still abound over the relevance and appropriateness of degrowth?

BIRTH OF A PROVOCATIVE SLOGAN

Degrowth was coined as a mere notion, but with the clear intent of reversing growth, in 1972, when sociologist and journalist André Gorz contributed to a debate organised by the Club du Nouvel Observateur in Paris. Gorz asked a profound question with respect to the just published and later highly influential Meadows report *The Limits to Growth*.[6] Was 'global equilibrium', he asked, 'compatible with the survival of the (capitalist) system?' given that Earth's balance required 'no-growth – or even degrowth – of material production'.[7]

Later in the 1970s, 'degrowth' was used several times, and mainly as a direct translation of 'decline', as in Nicholas Georgescu-Roegen's work on natural degradation in *The Entropy Law and the Economic Process* (1971). So, in 1979, when Jacques Grinevald and Ivo Rens translated four of Georgescu-Roegen's essays into English, they agreed to use 'degrowth' in translating the title *Demain la Décroissance: Entropie – Écologie – Économie* into *Tomorrow Degrowth: Entropy – Ecology – Economy*.[8] Subsequently, in the 1980s and 1990s, 'degrowth' appeared from time to time at conferences and in publications but most of the time as a synonym for 'decline', such as in the monthly magazine *S!lence* in a 1993 special issue on Georgescu-Roegen that was edited by Grinevald. As such, the word was really only used occasionally. Although used with great precision and intent, the response was hardly fireworks. However, at the beginning of the 2000s, all of this changed.

An Adbuster activist group in Lyon who feared the greenwashing and re-appropriation of the concept of 'sustainable development' by the capitalist system read Georgescu-Roegen and realised that '*décroissance*' might be a powerful semantic tool to radically question the limits of growth. That same year, in 2001, a group of intellectuals published on such themes in a special issue of the periodical *L'Écologiste*: *Unmake Development, Remake the World!*[9] This was followed by a colloquium of the same name, from 28 February to 3 March 2002, at the United Nations Educational, Scientific and Cultural Organization (UNESCO) in Paris, organised by La Ligne d'Horizon.[10] Consequently, these two groups got together to collaborate on a 2002 special issue of *S!lence*, a special issue that they called *Décroissance Soutenable et Conviviale – Sustainable and Convivial Degrowth*.[11]

Even if their first understanding of 'degrowth' was in response to Georgescu-Roegen's work and the need to decrease, to radically reduce, production and consumption, Adbuster activists Bruno Clémentin and Vincent Cheynet immediately saw in *'décroissance soutenable'* (sustainable degrowth) an alternative slogan to *'développement durable'* (sustainable development). Vincent Cheynet had been a marketing project manager with a keen eye for attention-grabbing slogans. Now the skills of promoting commodities for sale would be turned on their head in an effort which was anti-consumptionist and, indeed, more along the lines of decommodification. Meanwhile, members of the more academic and intellectual group were exploring the anthropological and cultural limits to growth, rapidly adding new dimensions to the emerging idea of degrowth.

The *S!lence* special issue included a contribution by degrowth pioneer Serge Latouche 'A bas le développement durable! Vive la décroissance conviviale!' ('Down with sustainable development! Long live convivial degrowth!'). Here, very clearly, the degrowth attack on growth was explicitly undermining the reformist concept of light-green 'development' and highlighted its anti-systemic direction. Latouche wrote: 'To survive or endure, it is urgent to organise *décroissance* ... it is not enough to moderate current trends, we must squarely escape development and economism'. Degrowth became a political force: 'Enacting *décroissance* means, in other words, to abandon the economic imaginary, that is the belief that more equals better.'[12]

In short, *décroissance* was a slogan born of radical anti-system critics who wanted to alert the world to the physical limits of growth and to question both the meaning of life and the imperialist dimensions of development. So much so that Gilbert Rist, author of *The History of Development: From Western Origins to Global Faith* (1996), would write of 'degrowth' that 'this neologism, was indeed an effective and genuine marketing coup which could have only be made by real professionals, even if we are all conscious about the ambiguity behind the term'.[13]

AN EFFECTIVE SEMANTIC TOOL?

A semantic tool enabling us to explode the concept and centrality of economic growth and question growth-associated addictions, 'degrowth' now became a tool for inviting in-depth debates on the unsustainability of infinite growth on a finite planet and to question whether growth was ever desirable. Even if criticising growth is not new, and its sabotage has been driven by others as well as degrowth advocates, 'growth' remains the dominant concept and simplistic, quasi-religious belief, in capitalist societies.

Growth is an omnipotent solution to all our problems – even, perversely, those problems that growth has caused – from unemployment to rising inequalities, from economic crises and public debt to environmental crises, energy scarcity and even starvation. Mainstream politicians, journalists, commentators and academics hail growth or its veritable namesake 'development' while the main goal of degrowth advocates is to attack the belief that more means better. In fact, Latouche has argued that we should speak about *a-growth*, in the same way as we speak about atheism, for it is a liberation from this belief of 'always more'.

Similarly, the intent of 'sustainable degrowth' was, first, to highlight the intellectual deception of the oxymoron 'sustainable development'; second, to avoid the risks of systemic re-appropriation; and third, to preserve the momentum of the rising awareness of, and discussion and debates around, environmental limits. Two decades later Cheynet's and Clémentin's concerns about the social power of greenwashing have been validated. Today everyone is encouraged to buy all kinds of familiar products mainly produced and distributed as they always have been but packaged anew with environmentally friendly advertising and narratives: sustainable, green, smart, good for ecosystems, 'bio', 'organic' and even 'fair', as in fair trade.

Using the term 'degrowth' protects against such dynamics. It is much more difficult to empty 'degrowth' of its clarity of meaning and its radical critique of the rationales imposed by capitalism, productivism and consumerism. Degrowth semantically challenges us to continually re-question all those drives and forces behind what we produce, how we produce things and for what uses. Using 'degrowth' protects advocates from linguistic distortion or co-option by capitalist forces and protects the movement from false and simplistic solutions to achieving environmental sustainability, such as green techno-fixes.

CHAPTER OUTLINE

Degrowth has spawned, and attracted, a variety of complementary concepts less provocative but more connotative, indeed 'federative', in their association. A significant conference where the degrowth movement made presentations, informed and engaged with the European Union (EU) Parliament (in Brussels) in September 2018 was nominally a 'post-growth' conference.[14] Given that the degrowth lexicon embraces terms such as conviviality, decolonisation, political ecology, socio-ecological economics, happy sobriety, voluntary simplicity, ecofeminism, municipalism, green new deal, transition, permaculture, prosperity without growth and autonomy, it brings the movements associated with them into a veritable convergence.

Degrowth is an invitation to go on the inevitably long journey of the decolonisation of our growth imaginaries, moving from cultural awareness to a systemic and material transformation changing our everyday practices. Degrowth insists on the deconstruction and re-evaluation of beliefs within, and relations between, capitalism and productivism, consumerism and materialism, development and the quasi-religion of economism, science and technology. In chapter 2 there is a discussion of the evolving matrix of complementary thoughts that have underscored the evolution of the idea of degrowth and follows how their integration with one another, that is, their 'articulation', ultimately questions the implicit and all-pervading belief in growth that dominates our contemporary reality. As such, the process of decolonising our imaginaries even challenges our concepts and daily practices around time, gender, death and democracy.

Beyond a set of thoughts deconstructing our dominant social and political structures, degrowth is an invitation to join a personal, collective, political, non-linear and heterogeneous journey that cannot be summarised in simple

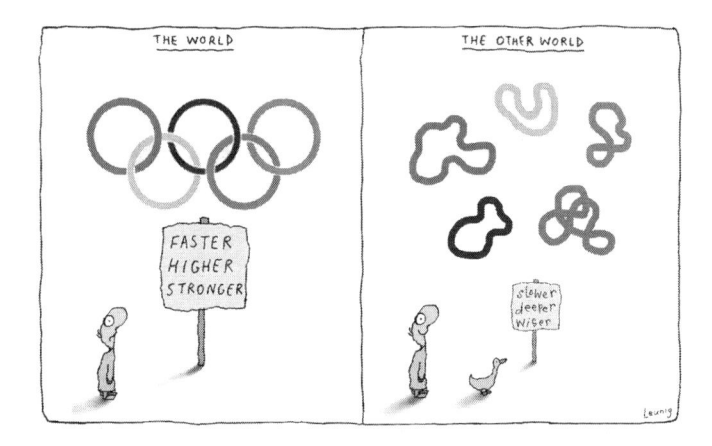

Figure 1.1 Two imaginaries
Cartoonist: Michael Leunig (Australia)

attractive slogans. Understanding degrowth requires time and iterative discussions to think deeply about the preconceptions, appearances and feelings that the word 'growth' creates. Degrowth calls for this intellectual, personal and collective effort. Sexy or not, federative or not, those in the Global North with reasonable incomes need to degrow to live within Earth's environmentally sustainable limits. Simultaneously, there is a need to degrow inequalities within all societies. Inequitable ways of life are not sustainable, generalisable, or even desirable – as indicated in Figure 1.1. Like it or not, the movement challenges us all to imagine and move towards a degrowth future in responsible, democratic and emancipatory ways. As such, in chapter 3 we explore the spheres in which activists within the degrowth movement practice: as individuals, collaboratively in cooperatives and collectives, in resistance and furthering the degrowth project.

The key debate goes beyond whether we *want* to pursue degrowth. The main question revolves around how we implement what we perceive of as the *inevitability* of degrowth. The rise of carbon emissions and biodiversity loss are symptoms of unsustainable forms of production and associated unsustainable lifestyles. Climate change is just the tip of the iceberg of our current environmental crises. The cracks in the current system are growing wider, with uncontrollable natural disasters and oppressive economic woes. *Just as we must cut carbon emissions, we must reduce overall consumption, and production.*

In short, the key question is: will degrowth be democratically chosen and implemented in sustainable, fair and convivial ways or imposed as an unfair, undemocratic and violent decline, even collapse? It is legitimate to ask: isn't the choice between degrowth or barbarism?[15] How has the degrowth movement moved from a slogan to that 'hard slog' of putting degrowth into practice, to start processes of new ways of living and being? In chapter 4 we explore the political dimension of degrowth. How might the movement best articulate pragmatic strategies of 'doing', experimenting with and prefiguring a degrowth society, engaging with parliamentary representation, collaborating with the union movement and, crucially, initiating non-violent direct resistance in terms of rapacious capitalist growth? We find that internal organisation and external relations are two sides of the same, essentially political, coin. Deepening the analysis, chapter 4 discusses political philosophy and analyses strategy to pre-empt the debates on the degrowth agenda, the 'degrowth project', which is the focus of chapter 5.

AMBIGUOUS, LIMITED AND LOST IN TRANSLATION?

A provocative anti-systemic slogan, degrowth has attracted criticism, even loathing and disgust, specifically as a way forward. Isn't degrowth just 'going back to the caves'? Isn't it a dangerous form of austerity? Why don't we just demand better forms of growth?

Even if its influence within political, intellectual and cultural debates has been significant and keeps increasing, those involved are eager for the degrowth movement to become more visible. Certainly, in 2020, degrowth cannot be said to have inspired an exponentially growing mass movement. Nor has it ever been as prominent as the Occupy movement was when it started in 2011. Long-time European proponents are impatient and disappointed that the potential of the deep philosophy and radical direction of degrowth has not been realised. Why hasn't the movement attracted more support? They have even wondered whether this indicates that degrowth is just a transitory tool, a simple demand or campaign, within a much broader transformation?

The key point is that, so far, degrowth has not been able to create, federate or comfortably federate within a large movement. But isn't this the situation for almost all progressive and emancipatory movements this century, especially since 9/11, Occupy and the alter-globalisation movements? We address reasons for this seemingly unnecessary sense of malaise and, more significantly, make clear and radical proposals for ways forward in chapter 5. Meanwhile, degrowth is unlikely to disappear as a social force until the decolonisation of our economistic growth imaginary – and addictions to more, better and bigger – is achieved.

More worrying are charges that degrowth is 'reactionary', that the use of the Latin negation *de* before a quantitative, economic term is contradictory. And, does not the very word 'degrowth' create the risk of falling into a binary trap of endless debate for or against growth? Advocates argue that degrowth must encourage a movement of focus from quantitative to qualitative perspectives, and aims to enable more complex and subtle debates. That's why a few still wonder if *a-growth* might be a more powerful and precise slogan.[16] Because it was developed specifically to directly hit its target, growth, it is a paradox that degrowth is subject to so much misunderstanding. Of course, to degrow for degrowth's sake would be as stupid as it is to grow simply for the sake of growth.

Often, we hear or read what seems to be a reasonable statement, such as: 'I agree with the ideas and the principles behind degrowth but not the term.' This objection was raised several times in the aforementioned EU Parliament Post-Growth Conference in 2018. It is a favourite statement made by politicians, economists and entrepreneurs who prefer to sweeten whatever they think is 'poison'. Some conference participants responded to the effect that it might be good that degrowth was an uncomfortable word to their ears, even better a thorn in their sides! Many in the movement consider that disliking 'degrowth' as a word is a key indicator that the person making the objection has more substantial misgivings, would prefer to hide from or trivialise the radical meaning of degrowth, its depth and holism.

Still, it is tiring to have to explain what degrowth is *not*, through denial and defence, rather than concentrate on a constructive discussion about *the real meaning of degrowth*. A word less open to disagreement might be preferable but 'a-growth', for instance, has never taken hold. There is a fear that 'a-growth' might meet the same fate as 'sustainable development', that is, it would be appropriated or co-opted. Many related and unrelated political slogans have faced misinterpretation and distortion, most often due to false associations made with their practices and practitioners. As they say, 'mud sticks'. 'Simple living' has been tainted by 'hippies', women's liberation has been altered to 'feminism', and 'socialism' has been adopted to make a clear distinction from failed real-life examples of state 'communism'. At the same time, there has been widespread resistance to such changes in nomenclature following the strategy of *owning* the name one is acrimoniously called, as in 'Yes, we are Queer and we celebrate it!'[17] In a somewhat similar way, for us in the movement, 'degrowth' sticks.

Décroissance spread from France to French-speaking regions like Belgium and Switzerland. First, it was translated into other Latin languages – in an intuitive way based on the logic of French, with the Latin negation '*de*' followed by '*croissance*', thus '*decrescita*' in Italian and '*decrecimiento*' in Spanish. By the time of the First International Degrowth Conference in Paris in 2008, it had been translated into English as 'Degrowth' but not without a prolonged debate. Indeed, the call for conference abstracts included both 'De-growth' and 'Degrowth' – which remained in simultaneous use until a consensus was found with the non-hyphenated and finally de-capitalised form.[18]

From the late 2000s, *décroissance* was translated into many other languages – as '*nemnövekedés*' in Hungarian (2011–) and '*odrast*' in South Slavic languages of former Yugoslavia (2013–). In Germany degrowth became *Postwachstum*, *Postwachstumsgesellschaft* or *Postwachstumsökonomie* given that 'linguistically it is not possible in German to construct a neologism in parallel to "de-growth"'.[19] However, this causes some confusion due to widespread English interpretations of 'postgrowth' as not necessarily anti-capitalist, let alone consistent with the principles of degrowth. What has made everything easier for English-speaking cultures is that other languages, such as Swedish, simply adopt or re-use the English word.

THE DEBATE IS STILL OPEN

This chapter set the rise of the degrowth movement in our global conjuncture, with global environmental crises and economic instability framing debates on the uncertain future of both our human species and the growth-driven capitalist system. We sketched the evolution of the 'degrowth' movement from a critical meeting between authentic Adbusters who wanted a foolproof name for their movement and intellectuals absorbed by scientific works focusing on physical and philosophical limits to growth. From a provocative slogan, degrowth has become a rich and multidimensional set of thoughts, theories and ideologies starting with the deconstruction of our growth imaginaries right through to the multi-faceted material challenges of emancipation.

There is no doubt that the term 'degrowth' has proved a useful platform for fruitful debate and experimentation. Yet it too has its limits and there is a general regret around the antipathy, dissonance and misunderstandings that the word 'degrowth' can evoke when people first encounter it, and even after some familiarity with the term. So, we have to ask: is such a provocation still a useful tool? Yes, for the last twenty years it has attracted rising interest and a growing movement and, most significantly, the word 'degrowth' has avoided co-option, proving both its pertinence and its utility.

CHAPTER 2

Decolonising Our Growth Imaginaries

From a provocative slogan, 'degrowth' has become a movement of activists and theorists who highlight the limits to growth. Degrowth means the transformation of society and the adoption of new models with qualitative, human-oriented and Earth-centred characteristics such as conviviality, autonomy and enjoyment of life, along with establishing principles consistent with ecofeminism and social and environmental justice. Most significantly for this chapter, degrowth theorists and activists aim to decolonise our growth imaginaries as well as identify principles and processes as pathways to transform beyond growth.

The degrowth movement burgeoned this century to spawn, adopt and critically adapt different types of proposals – from guaranteed minimum incomes and simple living to embracing post-development and post-capitalist visions. There have been heated discussions around what and how much needs to be changed, and which transformative ways offer the best outcomes. Is degrowth a strategy rather than a vision? Is degrowth commensurate with post-capitalism? Especially in Europe, where the movement began, there have been all kinds of experimentation with shared living, growing one's own food, working part-time and establishing ethical enterprises that concentrate on quality of life and ecosystems, applying ecological and social values (explored in chapter 3). Many of these practices draw on simple and collective ways of living by peoples in all parts of the world that have been practised for millennia.

This chapter sketches the evolution of a matrix of complementary thoughts that underscore degrowth and shows the articulation, or integration, of each theory with others. In short, degrowth is about questioning the implicit and all-pervading belief in growthism that dominates our contemporary reality. As Herman Daly wrote in 2019:

> We have many problems – poverty, unemployment, environmental destruction, climate change, financial instability, etc. – but only one solution for everything, namely economic growth. We believe that growth is the costless, win-win solution to all problems, or at least the necessary precondition for any solution. This is growthism. It now creates more problems than it solves.[1]

Degrowth invites a radical exploration of the sources of growthism in order to decolonise our imaginary. Later chapters discuss transformation, but this chapter examines some key intellectual sources of degrowth in order to drill down into those physical and philosophical limits to growth that the degrowth movement acknowledges, exposes and to which it responds.

GEORGESCU-ROEGEN'S ENTROPY LAW AND THE ECONOMIC PROCESS

The work of Nicholas Georgescu-Roegen has been highly influential within the degrowth network. He is often cited as a pioneer of degrowth. He emerged from the hard sciences but his deeply philosophical approach and intellectual journey both indicate that degrowth is remarkably interdisciplinary. Born in Romania in 1906, Georgescu-Roegen became a brilliant mathematician, receiving a scholarship to study at the Paris Institute of Statistics. There, he broadened his fields of study to encompass the philosophy of science, and, in 1930, defended his doctoral dissertation. In London, Georgescu-Roegen met the eminent English scholar Karl Pearson and this encounter extended his interests to biology, evolution theory and energy.

A scholarship to study at Harvard University with the outstanding Moravian-born United States (US) economist Joseph Schumpeter introduced him to economics and influential economists. However, Georgescu-Roegen turned down a position at Harvard's economic faculty to return to his home country and fulfil roles as a well-rounded expert in several fields. He applied his formidable mind working for various ministries and within Romanian politics, including in trade, peasant and agrarian reforms, and negotiations over the Soviet occupation.

After the Second World War, Georgescu-Roegen sought refuge from the Communist regime and worked temporarily at Harvard University. Finally, he ended up at Vanderbilt University in Nashville, where he died in

1994. In these final decades of his life he wrote his most well-known works radically criticising the foundations of economics: *The Entropy Law and the Economic Processes* (1971) and *Energy and Economic Myths* (1975). The journey of Georgescu-Roegen from statistics to biology and from physics to economics revealed thermodynamic proof that infinite growth on a finite planet is impossible.

Once he understood that the highly consumptive American way of life surpassed the capacity of the planet, he openly argued for 'decline'. His wide-ranging intellectual journey had connected deep interdisciplinary understandings of contemporary cultural, economic and environmental challenges to set the stage for the birth of degrowth. In the process, Georgescu-Roegen radically reformed neoclassical economic theory and models to include the second law of thermodynamics.

A few laws underlie thermodynamics as a branch of physics. The first two laws are relevant to degrowth. The first law, based on a conservation principle, states that energy is neither created nor destroyed in an isolated system but rather total energy remains constant. The universe is the ultimate isolated system. Energy can, however, change in form. The second law, which is based on a degradation principle and known as the entropy law, states that energy tends to dissipate; the entropy of the universe never decreases but only increases, gradually displaying more disorder. 'Entropy' is a thermodynamic property representing the unavailability of a system's thermal energy for conversion into mechanical work. Entropy is often seen as the degree of systemic disorder or randomness.

Georgescu-Roegen later developed a fourth law of thermodynamics, which is also significant to degrowth theory. His law states that material – not only energy – tends to degrade in the sense that it breaks down and its components separate. Georgescu-Roegen spent his last years developing a bioeconomics theory. He was a strong influence on the founders of the field of ecological economics, engaging in radical debates with one of them, his former student Herman Daly. He found Daly's concept of a 'steady state economy' spurious, the problem being that contemporary wealth, in terms of its use values, often depends on geological concentrations of high entropy materials (non-renewable energy and materials) that cannot be replaced in a useful time frame to keep up with reproduction at current levels.

Simultaneously, he warned about the risks of promoting so-called sustainable development, which he referred to as 'snake oil'.[2] Some degrowth proponents who share Georgescu-Roegen's radical critiques of sustainable development and question the limits of steady state economics try to continue his work on bioeconomics. Others, following Serge Latouche and radical critics of neoclassical economics and economics as a scientific approach, have proposed other ways of undercutting the quasi-religion of economics. Very scientific, often misunderstood and neglected, Georgescu-Roegen's work went far beyond physics, entropy and examining economic processes.

He made creative interventions of a philosophical nature, radically questioning the absurdity of Western thought, culture and the relatively recent growthism:

> We should cure ourselves of 'the circumdrome of the shaving machine' … which is to shave oneself faster so as to have more time to work on a machine that shaves faster so as to have more time to work on a machine that shaves still faster, and so on ad infinitum …

Instead, he counselled: 'We must come to realize that an important prerequisite of a good life is a substantial amount of leisure spent in an intelligent manner.'

Indeed, Georgescu-Roegen went on to argue for simplicity and frugality, for low-tech and local production, for liberation from extractivism and fossil energy, for solidarity and to stop wars – all strategies that feature in degrowth theory and practice today. He felt 'completely vindicated' that 'from a purely physical viewpoint, the economic process only transforms valuable natural resources (low entropy) into waste (high entropy)'. Yet:

> the puzzle of why such a process should go on is still with us. And it will remain a puzzle as long as we do not see that the true economic output of the economic process is not a material flow of waste, but an immaterial flux: the enjoyment of life.

Thus, Georgescu-Roegen drew strong philosophical conclusions from his study of the physical processes of bioeconomics. These conclusions did not so much transfer physical theories to social analyses and proposals but rather made a distinction between the material and the social. Most significantly, his work highlighted the material limits within which people needed to live. Simultaneously, he identified their qualitative potential.

GROWTH AND GROSS DOMESTIC PRODUCT

The concept of 'gross domestic product' (GDP) was created and used in the mid-1930s to measure the impact of the New Deal on the US economy. Subsequently, GDP would become the main indicator of the state of national economies after the 1944 Bretton Woods Conference. From a simple measure, which was criticised initially for its limits and risk of abuse, GDP would become the central indicator, a proxy for the success of a country, a policy or a government. More than that, it has become a veritable totem attracting almost religious fervour, a curiously specific quantity of an imprecise and changeable substance.

The danger of falsely magnifying and misconstruing the significance of GDP was even recognised by its progenitor, Simon Kuznets, who, in a 1962 article, wrote that 'distinctions must be kept in mind between quantity and quality of growth, between its costs and return, and between the short and long run'. In short, 'goals for "more" growth should specify more growth of what and for what'.[3]

Although often regarded in positive ways, GDP growth simply indicates a total *monetary* amount of production and services traded, disregarding any of the environmental and social implications of its components. Hence, a fire in a chemical factory is good for GDP growth. Meanwhile a very pleasant ride on a bike in the forest with friends and family is not even captured by GDP. Efforts have been made to correct such absurdity, for instance by creating metrics such as the Genuine Progress Indicator (GPI). However, the GDP maintains it status as the prime indicator of the state of society, blinding us to the key social and environmental challenges that we face as humans on planet Earth.

Even from a basic mathematical perspective monetary growth is absurd. To apply such an exponential function, say around 3 per cent growth per annum, the size of our economy would multiply by almost 20 times by the end of the century! Do we want such exponential growth? Often touted as directly correlated with well-being and welfare, many studies show that GDP growth does not necessarily mean more happiness. In contrast, the degrowth movement invites governments to give up their addiction to GDP growth, to turn from quantitative reductionism and fetishism to qualitative, human-centred and environmentally aware assessments of the costs of inputs and outputs – including waste – from our productive activities.

FROM 'PEAK EVERYTHING' TO 'COLLAPSOLOGY'

Capitalism is based on exponential growth. Since the first Industrial Revolution, the beginning of industrialisation and use of fossil energy – initially coal and later oil and gas – a 'great acceleration' of activity eroded the regenerative capacity of our planet. From energy consumption to demography, from environmental footprints to the destruction of fertile soil, from the number of cars per capita to levels of meat consumption, all these curves grew exponentially and have exploded since the Second World War and dependency on oil.

Oil, the most efficient, easy to transport and concentrated source of energy, surrounds us. Coal, oil and gas continued to account for more than 80 per cent of global primary energy demand and electricity generation this century, from 2000 to 2017.[4] They are used as materials in thousands of products, from asphalt to fertilisers, plastics, paints and clothing. We consume oil. We require oil to travel and to communicate. Digital technologies currently account for 3.7 per cent global greenhouse gas emissions – well over those of the civil airline industry (2 per cent) – having increased from 2.5 per cent since 2013, and information and communication technology energy use is rising by 9 per cent per annum.[5] Consequently, within a few decades we will have burnt what took tens or even hundreds of millions of years to accumulate in plants from solar energy.

In short, the vast majority of our primary energy has come from a limited stock which was developed through natural processes of long duration. Now we are reaching peak production of this illusory 'free energy' (Figure 2.1). 'Peak oil' does not refer to a total depletion of oil reserves but rather a level of exploitation beyond which the recovery of further reserves is not economic. Peak conventional oil was reached in the mid-2000s, only to be compensated for by exploiting unconventional oil, such as shale-oil or bituminous sand, which are extremely polluting. Oil exploitation has had high environmental costs. Fossil fuel dependency is directly responsible for climate change and, indirectly, oil exploitation is responsible for a host of other forms of environmental damage associated with a range of processes and products. These include oil spills and contamination over land and in oceans, ground water and surface waterways.

Figure 2.1 Energy descent
Artist: Patrick Jones (Australia).

Substitution with so-called renewable energies is an illusion because such 'alternatives' rely to varying extents on oil for extraction, production, transport, maintenance and end of life management. These alternatives are reliant on complex infrastructures for their establishment and operation. Moreover, they are dependent on limited stocks of other resources, such as metals and rare earth minerals, all with numerous negative environmental implications.

Nevertheless, in the conventional press and academies, there is still talk about 'green', 'smart', 'sustainable' and 'inclusive' growth, as if these goals were achievable. Growth appears in all the official documents and goals of influential institutions, such as the European Commission and the Organisation for Economic Cooperation and Development. 'Sustainable development', as in the 17 goals of the United Nations, is about salvaging growth by maintaining the illusion that it is possible both to keep growing and to reduce our environmental impacts. Very often, the solution touted is technological, with myths of 'dematerialisation' and 'smart technologies', what economists call 'decoupling'.

Decoupling is the notion that negative environmental implications of production and consumption can be addressed separately from growing the economy. In short, we can achieve capitalist growth without environmental destruction, or with minimal environmental destruction. The problem is that decoupling is highly unlikely to eventuate, as argued in this mid-2019 report:

> The conclusion is both overwhelmingly clear and sobering: not only is there no empirical evidence supporting the existence of a decoupling of economic growth from environmental pressures on anywhere near the scale needed to deal with environmental breakdown, but also, and perhaps more importantly, such decoupling appears unlikely to happen in the future.[6]

Georgescu-Roegen's works about the physical limits to growth now exist alongside many other works on the environmental challenges of the thermo-industrial age in which current generations live: climate change, biodiversity loss, acidification of oceans, destruction of the fertility of our soils, chemical and atmospheric pollution, from peak oil to 'peak everything'.[7] However, degrowth networks, particularly in France, have criticised the simple promotion and circulation of this type of information about rising dangers and impending catastrophes as reductionist and counterproductive. In other words, accumulated together in overwhelming forms of 'collapsology', such accounts end in disabling alarm and panic.

While it is necessary to keep in mind the frightening dimensions of growth, degrowth advocates prefer to initiate emancipating and democratic debates on a desirable transition away from growth. Degrowth advocates and activists are keen to invent, experiment with and implement new prefigurative models based on other principles. Degrowth forums address the philosophical dimensions of degrowth. Of course we know that over-consumption is neither sustainable nor generalisable because of the physical and environmental limits of the earth. But, as Earthlings, we also have to ask ourselves: Is unending consumption desirable? Even if infinite growth were possible, would infinite growth be meaningful?

In summary, with respect to physical limits, we refer to a cautionary tale about the snail told by another great contributor to degrowth theory, the philosopher Ivan Illich:

The snail constructs the delicate architecture of its shell by adding ever increasing spirals one after the other, but then it abruptly stops and winds back in the reverse direction. In fact, just one additional larger spiral would make the shell sixteen times bigger. Instead of being beneficial, it would overload the snail. Any increase in the snail's productivity would only be used to offset the difficulties created by the enlargement of the shell beyond its preordained limits. Once the limit to increasing spiral size has been reached, the problems of excessive growth multiply exponentially, while the snail's biological capability, in the best of cases, can only show linear growth and increase arithmetically.[8]

Because we need to slow down, and even more because of the influence of this tale within the movement, today the snail is a familiar symbol of degrowth.

FROM DEVELOPMENT TO DECOLONISING OUR IMAGINARIES

Even as the illusion of decoupling through innovative, green and smart technologies becomes more widespread, there is increasing acknowledgement of the physical limits to growth. Writing in 2020, we find a consensus in the scientific literature, in particular around observed and reported acceleration of climate change and biodiversity loss. As scientific evidence is synthesised and highlights various and numerous physical limits to growth, degrowth also makes more sense in terms of an appropriate and systemic socio-cultural response. Degrowth illuminates the socio-cultural limits to growth.

Chapter 1 introduced degrowth as a fortuitous meeting of two groups. On the one hand, there were Adbusters who feared that 'sustainable development' would be greenwashed. On the other hand, there were academics who argued that we should totally 'unmake development' and, instead, 'remake the world'. We have argued that 'development' – like 'growth' – is a loaded word. Indeed, Serge Latouche has pointed out that in an attempt to translate 'development' into Eton – aka Ìtón, a Buntu language used in Cameroon – the closest approximation was 'the white man illusion'![9]

Questioning the socio-cultural limits to growth and development opens the door to emancipatory debates about the meaning of life. To unpack 'development' means deconstructing belief in material 'progress' as either inevitable or natural. We have to understand the origins of the term and theory of 'development', how development became a worldwide hegemonic ideology, a postcolonial and imperialistic cultural tool.

In his 1949 inaugural address, then US President Harry Truman patronisingly distinguished 'developed' from 'underdeveloped' countries in terms of the former helping the latter to follow down the same road.[10] The colonial pillaging of effectively 'under-Westernised' countries persisted due to what alter-globalist Animata Traore (Mali's ex-Minister of Culture and Tourism) has called a 'rape of their imaginary'.[11] Illuminated in works such as Uruguayan Eduardo Galeano's *Open Veins of Latin America: Five Centuries of the Pillage of a Continent*, the extraction and exploitation of resources from Africa, Asia and Latin America was secured and organised to support incredible economic progress in Europe. This is how the period 1945 to 1975 became known generally as the 'Golden Age of Capitalism', 'the Glorious Thirty' years in France, the post-Second World War economic boom for most countries of the Global North.[12]

Even if development is frequently referred to in one-dimensional ways as positive and natural, holistic perspectives have clearly shown that development is neither necessary nor desirable.[13] The Iranian diplomat and post-development theorist Majid Rahnema's book *When Misery Hunts Poverty* pointed out the sharp contrast between, on the one hand, the forced misery of degrading work within the imposed scarcity of capitalism and, on the other hand, living modestly in voluntary simplicity.[14] The environmental, human and cultural consequences of globalisation continue to be devastating. But there are alternatives. Degrowth invites you to take account of capitalism's imperial dimensions and challenge the prevailing system's hidden human and environmental debts which include exploited and destroyed ecosystems and peoples.[15] By 'decolonising our growth imaginaries' we can re-imagine genuinely progressive models of society.

THE COUNTER-PRODUCTIVITY THRESHOLD

Illich's cautionary tale about the snail – the snail's options and decisions – warns that striving for more, bigger and stronger, sooner or later reaches a 'counter-productivity threshold'. Illich provided an example of the counter-productivity threshold by calculating the average 'speed' of a car in a more holistic way than is normally done. Illich took into account how many hours one might need to work to finance purchasing and operating a car; to cover the costs of the public infrastructure that driving cars requires; and the negative impacts of cars in terms of both their production, operation and waste and the results of car accidents. Reframing the way one might measure the time that a car 'saves' in these ways, Illich reduced the car from seeming to be a very powerful device for travelling 100 km per hour to a device that simply achieves an average speed of 5 km per hour.[16] In short, Illich's calculations indicated that it is as quick to go on foot and even quicker by bicycle!

Another salutary aspect of such calculations is the realisation that, if we run this scenario for those on higher incomes, then they are more quickly able to pay for all the costs involved, so the car is indeed a 'quicker' proposition than it is for someone on a low income! This lesson shows how income inequalities alter the real everyday experience of the social costs of any form of technology. Politically, this translates into a stronger tendency for managers and owners on high incomes to value technologies because they have a lesser impact on their hip pocket than those on lower incomes. This is an added reason for a concern with reducing inequalities of income, which are addressed in detail in chapter 5.

Illich's concept of the counter-productivity threshold can be applied to almost everything we do. For instance, one might need to weigh up buying a cheaper dwelling far from the city centre where one works with evaluating the time and monetary costs of various forms of travelling from home to work and back, all forms of transport involving a variety of infrastructure, social and environmental impacts. This example also raises lifestyle matters: how much do we enjoy or dislike the experience of driving, going by train, bus or tram to work, which can take up extended periods of time in two directions five times a week?

Matters of lifestyle relate to our enjoyment, or not, of experiences. By way of another example, using emails enables one to instantly communicate as much as one wants with colleagues and friends around the world. Ideally, this incredible, apparently effective, tool would liberate one, delivering freer communication and freeing up one's time. No need to go to the post office, wait in a queue, or walk to another building for a face-to-face discussion. Yet, many people end up stressed by too many emails and feel under strong social pressure to answer them quickly. Degrowth invites you to rethink your values and relations with respect to socio-cultural impacts from technological developments, in short to re-evaluate your use of everything.

THE REBOUND EFFECT

This discussion of the concept 'counter-productivity threshold' also raises the question of the 'rebound effect', whereby a technical solution is seen, ultimately, to transfer a problem elsewhere or even increase it. For instance, to reduce the environmental impacts of the standard car, we improved its design and created more effective engines so that cars became lighter and required less resources to be used in making them. However, given our system of production for trade, the consequence was not only more sales of cheaper cars but also cars were promoted with more optional extras. This technical advance meant that people could buy bigger cars fitted with more air conditioning and other gadgets.

Beyond functionality, such features have made certain models, and add-ons, aspirational status symbols. For instance, in the first nine months of 2019, Australians bought 50 per cent more new sports utility vehicles (SUVs) than they did standard passenger cars: 45 and 30 per cent respectively.[17] Moreover, more and bigger vehicles result in more traffic, more infrastructure for parking and roads for travelling on. More traffic often translates to slowing the speed at which one can get to various places. So, instead of reducing environmental impacts, the lighter-car approach inadvertently increased the total environmental impacts of road vehicles. This is what is meant by the term 'rebound effect'.

TECHNICAL TOTALITARIANISM AND THE BANALITY OF EVIL

Last and not least in a string of useful concepts related to degrowth thinking is what Jacques Ellul and Bernard Charbonneau have called 'technician totalitarianism'.[18] Again our system of production for trade through private firms competing with one another means that, say 'smart' phones are constructed and sold without any obligation, or even possibility, to measure all their cultural, social and psychological impacts and implications in terms of environmental and human exploitation. Moreover, people are increasingly forced to use a smartphone because governments and businesses assume everyone has one. Therefore, taxation and welfare and general services delivered to citizens often require use of such devices. At the same time few people are able to repair smartphones and respond to advertisements for a barrage of models with new and improved features, thus becoming victims of 'planned obsolescence' (whereby perfectly functional objects are discarded simply because they are unfashionable or outmoded; see Figure 2.2). As Serge Latouche pointed out: 'Advertisements create the desire to consume, credit provides the possibility to do it, and planned obsolescence renews its necessity.'[19]

Figure 2.2 Brands aren't your friends
Source: Special Patrol Group (UK)

While presented as liberating, algorithms decide what we see and receive, suggest what we do, where we go and how we live. Social networks, supposedly here to connect us to the world, enclose us in bubbles infiltrated with advertising that exists to optimise profits. From the appalling working conditions of miners extracting raw materials in the Congo and the heavy metal pollution from old smartphones waste disposed of in Ghana to our data becoming used for business, we find Hannah Arendt's 'banality of evil' arising in our increasing dependence on such heteronomous technologies.[20]

One could apply a similar logic and conduct thorough research in terms of most high-tech devices that are used every day. But, looking in the opposite direction, one can often find that small and simple are both beautiful. Degrowth proponents invite people to re-appropriate the management of technology and exercise direct democratic control, which can only be achieved by sharing otherwise-patented open-source knowledge, and by re-evaluating what people want, and how people want to produce such goods and services, that is, whether, how and how much we, as a collective, as communities, as workers, *want* to use high-tech or low-tech.

GROWTH, LIFE, HEALTH AND WORK

The French heterodox economist Bernard Maris – who was killed in the infamous *Charlie Hebdo* attacks in Paris in January 2015 – analysed capitalism and people's addiction to growth from a psychoanalytical viewpoint. He theorised that the compulsion to accumulate was driven by a fear of death. He argued that people unconsciously accumulated to stave off death and even in an effort to buy immortality. This is perverse: our growth-addicted society, in denial and unable to admit to and to deal with limits, pushes us into a morbid race of always-more. People work more to earn more and, finally, realise that they have missed out on what was most important – to experience meaningful and enjoyable lives (see Box 2.1).

Box 2.1 The tourist and the fisherman

It's a bay on a coast in West Europe. A tourist, snapping shots of the idyllic scene, wakes a fisherman lying asleep in his tiny boat. The dazed man feels around for his cigarettes but the tourist hands him one of his own, and offers to light it for him.

'It's a wonderful day for fishing,' he observes. 'When will you go out?'

'I won't,' replies the fisherman.

'Why not? You're sure to catch a lot! You're sick?'

'I feel wonderful,' answers the fisherman.

'So, why not go out and fish?'

'I've been out. I got plenty of fish: four lobsters and over twenty mackerel. Enough for a few days. Want a cigarette?'

The tourist takes the fisherman's cigarette, puts down his camera and stares at the fisherman. 'Why, you could go out again a few times today, and catch a lot more! From now on, you could go out a few times every day. In no time you'd save enough money for a motor, a year or two later you could buy a bigger boat. Then, once you were bringing in fish to fill a couple of cutters, you could build cold storage, a smokehouse and a pickling factory. Then you could supervise your workers by helicopter and get a licence for all the sea life in the bay. You could open a restaurant and you

▶

> could even wholesale lobster to Paris! What opportunities you have here.'
>
> The fisherman is unmoved.
>
> The tourist, with a whimper, 'But you could...'
>
> 'Could what?' asks the fisherman.
>
> The tourist is overwhelmed with pity, frustration and disbelief. Yet he puts aside his irritation to speak slowly to the fisherman: 'If you set yourself up like that, you would have so much leisure time. Why, you could wander down here, and lie on a deckchair in the sun, without a care in the world.'
>
> 'That's what I do when tourists don't bother me,' said the fisherman.
>
> The tourist walks away, shaking his head. He only has a short time off work. Soon he will return to his busy life in the city.
>
> Meanwhile the fisherman, with his simple life, will continue to work a little, relax a lot and enjoy the coast.
>
> Source: A summary retelling of Heinrich Böll's well circulated story, 'Anecdote concerning the lowering of productivity', translated by Leila Vennewitz, in *The Stories of Heinrich Böll*, 1st edn, New York: Knopf, 1986, pp. 628–30.

Degrowth is an invitation to, on the one hand, acknowledge environmental limits as a strict condition of any living species, any individual and any society. On the other hand, degrowth encourages a focus on key questions related to the meaning and enjoyment of life. Health and work are cases in point.

Questions of limits are easily apparent in our experiences of good health and in its maintenance. In France, certain medicines ultimately paid for by public health schemes are, at best, placebos. At worst, numerous medicines have serious, even life-threatening, side effects. Moreover, the average French person spends a large part of their health budget in the last year of their lives.[21] Isn't this approach to health symptomatic of emotional immaturity in that it indicates a denial of mortal limits? Playing at being Doctor Frankenstein, some people spend more and more on efforts to realise 'transhumanism' in the mere hope of upgrading their capacities instead of simply enjoying free time and good times with friends or enjoying their own company in contemplative ways. Following this train of thought, degrowth theorists – such as Ivan Illich in his *Medical Nemesis: The Expropriation of Health* (1975) and other philosophies and spiritual cautions about limits – invite everyone to re-evaluate what is really important to each of them.[22] By identifying what is significant one can, ultimately, contribute to a collective definition of our society's *socio-cultural limits*.

Meanwhile, many people experience life as rats in a wheel, running quicker and quicker to earn an income to be able to buy more stuff that they don't even need. They work to finance their car; they need a car to go to work! In 1880 the revolutionary Karl Marx's son-in-law Paul Lafargue published the first draft in French of what would become a wonderfully provocative book, which was translated as *The Right to be Lazy* (1883). A qualified and experienced doctor, Lafargue gave up practising medicine after losing all three of his children, reasoning that political change would be more effective in saving as well as improving lives. While his father-in-law had argued for the abolition of wages, Lafargue intervened for the 'right' to be lazy. Both objections to capitalist work have been swept aside by 'right to work' arguments in workers' movements and union activities.

The degrowth movement is inspired by those who challenge the practice of working for money and for profits. Degrowth advocates and activists call for a revival of debates around purposeful work, work that workers deem useful, work to fulfil basic needs, neither more nor less. André Gorz, who was influenced by Ivan Illich, resuscitated such discussions in 1980, in his classic work *Farewell to the Working Class* (*Adieu au Prolétariat*).[23] He challenged workers' movements to radically rethink salaried work in favour of activities that were chosen rather than forced. He argued for deconstructing the centrality of paid work, which is characteristic of capitalism. Gorz pointed out that work had its limits at both ends of the capitalist spectrum with exploitation at one end and, at the other, unemployment. The rise of precarious work, the gig economy and the 'precariat' magnifies these issues of socio-economic limits even further.

Mainstream economists and politicians iterate 'without growth, more unemployment'. But the causal connection is unclear, and especially tenuous when there is a widespread threat of future automation leading to mass unemployment. According to Éloi Laurent in *Measuring Tomorrow* (2017), the US GDP grew by 20 per cent in the 2000s while job creation actually decreased by 1 per cent. When growth does increase employment opportunities then it often means that stressful, exploitative, meaningless, useless, precarious and counter-productive work is on the rise. David Graeber has asked why people want 'bullshit jobs' that are boring and burn them out? The kind of work capitalism offers is the other side of the coin from growth for growth's sake.[24]

Moreover, a New Economics Foundation study found that very often, the more harmful a job is for society, the higher it is paid.[25] Contrast the low salary of a cleaner in a hospital, who performs valuable and burdensome tasks necessary to maintain people's good health, with the high salary of a financial adviser recommending legal tax evasions impacting negatively on health infrastructure. Clearly, the system is illogical not just inequitable. Consequently, in autumn 2018, 30,000 elite French university students signed a manifesto stating that they refused to serve a system based on GDP growth, a system which was simultaneously responsible for environmental and social disasters.[26]

So, how might we transform from a society centred around salaried work in jobs that people find objectionable, to a freer, more democratic society where people can direct their efforts to meaningful and emancipatory activities? Is there an escape from the barbarity of modernity with its senseless divisions of labour? How might a society be created in which everyone might enjoy a diversity of tasks every single day – a mixture of manual, intellectual, political, cultural and artistic, collaborative and contemplative tasks?

DEGROWTH IN INEQUITY: AUTONOMY, SELF-LIMITATION AND DIRECT DEMOCRACY

One of the degrowth movement's key slogans states that the first principle and stage of degrowth should be a degrowth in inequalities. As highlighted by economist Thomas Piketty in his bestseller *Capital in the Twenty-first Century* (2014), the last few decades have been characterised by exponential growth in inequalities.[27] It is becoming less credible to argue to that, thanks to growth, sooner or later, capitalism will profit everybody. The trickle-down theory states that, as the rich get richer, the economy grows and there is always extra to redistribute to the poor. Since the global financial crisis of 2008 in particular, we have seen recession, depression, austerity and negative interest rates in many countries. The trickle-down theory has proven to be a myth: many studies show the contrary to be the case.

A 2019 Oxfam annual report on inequalities revealed that just 26 people owned as much as the poorer half of the world's population.[28] These inequalities have implications for all aspects of everyday life. By way of an example, an Oxfam media briefing at the time of the intergovernmental negotiations on climate action, the 2015 United Nations Climate Change Conference, the 21st Conference of the Parties (COP 21), in Paris, estimated the inequity of drivers and victims of climate change. The poorest half of the global population – some 3.5 billion living in regions vulnerable to the effects of climate change – only cause around 10 per cent of total global emissions associated with individual consumption. In contrast, the richest 10 per cent are responsible for around 50 per cent of such emissions; the average ecological footprint of the wealthiest 1 per cent globally might well be 175 times that of the poorest 10 per cent.[29]

It is very clear, then, that deconstructing the myth of growth demands that we re-imagine our future in terms of challenging private property, sharing wealth and radical redistribution. In fact, many anthropologists and historians have contributed to degrowth theory in terms of critiques which can contribute to decolonising the growth imaginary. Karl Polanyi, in particular his work *The Great Transformation*, has been indispensable.[30] Polanyi showed that 'capitalism' is as much a political construct as its characteristic political and policy packaging as 'development'. While inviting us to explore other societies, Polanyi argues that in non-capitalist market societies markets were less central and that exchange values (prices) were encapsulated and counter-balanced by other values: political, cultural, social, traditional and spiritual. He analysed tragic outcomes of the twentieth century, such as fascism, in terms of the dominating central 'free' market in Europe.

Economic historians like Polanyi, and anthropologists such as Marcel Mauss, have challenged stereotypical theories of *Homo economicus* and reductionist neoliberal views of human practices claiming that individuals are singularly driven by personal interest. In this context, Serge Latouche refers to the 'invention of economy' or economicism, as if economics were a religion. As the 'law of the instrument' has it, with a hammer in your hands everything looks like a nail and, today, our hammer is the economy.

Following such radical critiques of capitalism, its unfair and violent construction via expropriation in what is now the Global South and broad-scale cultural hegemony, degrowth calls on people to decolonise their *Homo economicus* imaginary and to re-embed a shared economy. But, how can we all put the economy back into its rightful place, as a tool of democratic governance? Don't we all need to rethink private property and, instead, establish sharing and commoning?

Decolonising our collective social imaginary, freeing our socio-political minds and exercising direct democracy involves a radical critique of the educational system. Here helpful thinkers include Ivan Illich in works such as *Deschooling Society* (1971) and Cornelius Castoriadis, whose concept of autonomy challenges the roles of citizens within democracy. These authors' perspectives suggest framing democracy in cultural ways, in terms of daily practices, rather than in institutional ways. This requires re-appropriating and redefining everyday practices whether they be related to technology, production, justice or economic institutions. It involves deconstructing dominating structures and ideologies such as patriarchy and capitalism. Simultaneously, there is a need to construct social relationships and processes for every citizen to participate directly as an individual and collectively in determining societal laws.

FROM THE DECOLONISATION OF OUR IMAGINARY TO A TRANSITIONARY PROJECT?

In this chapter we could have quoted many more thinkers and schools of thought because degrowth draws on a multidimensional matrix of ideas from many cultures and periods of history. On the one hand, degrowth is an invitation to deconstruct certain hegemonic beliefs in, and myths around, say, development, how we construe work and institutions such as patriarchy. By scrutinising the social construction of what Latouche has referred to as 'toxic concepts', we can liberate our minds and take steps to construct fresh virtuous paths towards new models of society based on other values.[31] Beyond a decolonisation of our imaginary, degrowth is about co-constructing 'a vocabulary for a new era'.[32]

Degrowth is neither a dogma nor a reductionist concept. Rather, degrowth advocates identifying various ways to address contemporary challenges, discovering both potential within, and tensions between approaches that are found to be persuasive and potent. Degrowth is a set of thoughts aimed at shaking people's belief in growthism. In its place, degrowth offers a platform for debates and convergence, for re-appropriating hope through inventing, discussing, experimenting with and implementing democratic and peaceful transitions. Transitions to new models of society based on concepts unpacked in the next chapters, namely: the enjoyment of life, open relocalisation, sustainability, frugal abundance, conviviality and autonomy.

CHAPTER 3
Degrowth in Practice

Degrowth has attracted support as a multidimensional and influential contemporary movement that is both human-centred and critically concerned about planetary ecosystems. Sometimes misinterpreted by newcomers as a return to a primitive past, in reality the degrowth movement is very future-oriented – attempting to map ways out of hyper-consumption, inequality, weak democracy and the environmental crises caused by growth-driven capitalism.

In organisational terms degrowth is best described as a *decentralised*, *multidimensional* and *open* network, which is currently stimulating fruitful debates within existing and emerging movements, academic and political circles, and regional cultures. In this chapter we explore, in thematic and indicative ways, the degrowth movement's disruptive and enriching relationships with other movements with similar concerns.

Key questions for the movement have revolved around the extent to which degrowth transformations either simply require dramatic and extensive capitalist reforms or, instead, holistic system change to post-capitalism. How have degrowth and post-socialism in Eastern Europe converged, conflicted and informed one another? How has degrowth as a concept challenged socialists and been absorbed by ecosocialist thought? How far have other post-capitalist currents entertained, and integrated, degrowth as a strategy for transition? Is degrowth more than either a campaign or potential central demand of a series of associated campaigns? Is it, or can it be, an independent movement? Given that all such questions are open to contemplation, experimentation and debate, supporters and advocates take up varying positions and individuals adopt distinct or even uncertain positions over time sometimes due to changing circumstances.

Given the strong tendency to cluster, compound and dilute the meaning and force of degrowth as a demand in and of itself, we clarify its role synthetically illustrating its many shared goals and distinctions with respect to other movements. Local specificities and historic and cultural tensions impact on degrowth developments. Degrowth disrupts, engages with and supports other movements with similar approaches, including the transition, Extinction Rebellion and Occupy movements, as well as aligning to some extent with the transdisciplinary field of ecological economics.

The chapter focuses on activism and activists, revealing the diverse ways that activists have pursued degrowth by living simply; experimenting with alternative technologies and techniques for living and self-provisioning; forming political squats and social centres; campaigning against mega-infrastructure developments; experimenting with alternative currencies and non-monetary economies; developing action-based experiential methods and methodologies; imagining degrowth futures and discussing a degrowth agenda to 'get there', that is, working out strategies to implement a holistic degrowth project.

Key shared concepts are identified around 'spheres' of action and reflection emerging from degrowth activism. You will notice connections and contradictions between practices central to this chapter and theories outlined in chapter 2. A generic issue clear to practitioners is the ease with which abstract theory lays out visionary imaginaries without taking into account the very real barriers to achieving such transformation (issues detailed in chapter 4). At the same time, it is remarkable that the ideas and work of theorists of the late twentieth century remain so relevant, even as the social, political, economic and environmental crises have deepened.

Degrowth movements in specific locations develop certain defining characteristics. There are practices common to urban degrowth activists, such as squatting, supporting public transport and cycling, and developing relationships with food producers on the peri-urban or rural fringes of cities in arrangements such as community supported agriculture. In contrast, rural degrowth advocates can, arguably, more readily establish modest dwellings in ways known as do-it-yourself ('DIY') and do-it-ourselves ('DIO') – becoming substantially self-provisioning with respect to food and relying on slow and novel forms of transport.

A scholarly challenge is made to economists and ecologists to become more genuinely transdisciplinary and action-oriented, two directions which are already cutting-edge developments in several relevant social science fields, such as political ecology, disaster studies and action research. Such efforts to apply appropriate techniques for investigation and analysis of social movements and activism are essentially political in nature. As with climate change, science can never disarticulate itself from real-world economic and political implications.

A DECENTRALISED, MULTIDIMENSIONAL AND OPEN NETWORK

The concept of degrowth has certainly gained traction in the early twenty-first century as a provocative slogan derived from various strands of thinking. However, this interest and support has failed to evolve – so far – as a typical organised movement with a discernible formal structure. No international organisation or party has been established. Some attempts have been made to develop such a formal organisation. In assemblies within international degrowth conferences that attract thousands of participants, proposals supporting greater formality are often raised and fiercely debated.

Typically, certain spokespeople will claim that success relies on a clear and formally constituted organisation. However, many supporters insist that it seems more natural and realistic to assume and prolong the apparently 'unorganised' form of Occupy and other such movements of the twenty-first century. The devolution of the French Degrowth Party – Le Parti Pour La Décroissance (PPLD) – is instructive. As the articles of association section of its site explains, PPLD was established as a party that is no more. Playing with '*parti*', which means 'going to' as well as 'party', it retained its acronym but has been renamed 'Parti-e-s Pour La Décroissance' understood as 'on the way to degrowth'.[1]

Thus degrowth has organically maintained an existence common to grassroots movements of our times: an open network of individuals, collectives, projects, platforms and gatherings. This dynamic and mobile open network that degrowth has developed is analysed here. Initially the rise of horizontal organising as a feature of contemporary, especially radical, movements is discussed. Then the degrowth movement's four characteristic spheres of action are explained by teasing out typical expressions and activities in each sphere as well as identifying relationships between each sphere. These focal points of our analysis and spheres of action are the individual sphere, the collective sphere, the sphere of resistance, and the sphere of the degrowth project. The degrowth movement itself could well be mapped as a web of networks within and aside from such spheres.

NON-HIERARCHICAL ORGANISING

Since the 1960s, mass movements have evolved new forms of organising using techniques of direct action and non-violence, relating as equals in collective deliberation. The values underlying these developments are antithetical to the still mainstream model of capitalist bosses and managers; bureaucratic union officials and organisers; party politicians, representatives and spokespeople – all of whom lead and speak on behalf of a mass of institutionalised followers. A popular position that has evolved is explained in an interview with US activist-scholar Charlie Post in the radical UK *Salvage* journal:

> While I am currently not a member of an organization, my politics remain those of revolutionary socialism from below – a commitment to a revolutionary transformation of society, led by the working class (in all of its racial and gender diversity) which will establish a democratic collectivist order, no matter how distant the prospect. This perspective shapes my day to day activity in this clearly non-revolutionary period – struggling for immediate demands/reforms through organization and struggles that do not rely on elections, capitalist legality or the forces of official reformism (the labour officialdom, middle class liberal leaderships of communities of color, women, queer folk, etc.).[2]

Non-hierarchical forms of organisation – often interpreted and referred to as 'unorganised' by those with triumphal confidence in hierarchical methods – highlight the value of every person's understanding of the current circumstances, of each person's right to be heard, to express an opinion and to act according to their beliefs, that is, to be respected and treated equally in everyday practice. If such organisation evolved in everyday resistance and protest action, it is also viewed as prefigurative. Prefigurative of a community-oriented form of politics necessary not only to combat but also to go beyond contemporary hierarchies that facilitate the reproduction of patriarchal and racial oppression, authoritarian exploitation, confrontation and a military approach to people and planet.

These values and associated forms of organising have existed in certain times and places for centuries – as highlighted in the Paris Commune.[3] But, over the last 50 years, non-hierarchical organisation has gradually become common practice within radical elements of left-wing movements. Within the labour movement this thrust is expressed in a radical form of autonomous cooperatives, particularly well developed in *horizontalidad* in the early 2000s in Argentina.[4] By the early 2010s, surfacing in various cultural clothes around the world, the politics of horizontal organisation became front-page news with Occupy, the 15-M movement in Spain, young protesters rioting in Greece, student uprisings in Quebec and the Arab Spring.

Neither a reactive nor informal mode of organising, the assembly is the cornerstone of face-to-face organisation typical of transformative politicking. By way of an example, in the early 1970s in Melbourne (Australia) the women's liberation movement had an assembly every Saturday afternoon in a warehouse space. Any woman could attend. At the start of each meeting a facilitator and a note taker were elected and an agenda formed from everyone's input. Whoever was there made decisions in as consensual a way as possible; there was no president, secretary or president. In a daily drop-in centre women were rostered to act as contact points recording their activities and interactions through a diary and phone messages. There was no formal spokesperson. The media had to speak with whoever was on duty who would either pass them on to a specific liberationist or speak themselves for the movement but, in both cases, only as 'a' representative.

Another essential form of non-hierarchical organising is working groups, which might be permanent or short-term. They evolve to address a task or area of concern.

They are open to anyone to join and knowledge and skills are freely shared. Typically, working groups have a rotating contact point and coordinator responsible for briefing newcomers and reporting to assemblies. The roles of a working group are subject to assembly decision-making but they advise and take part in such decision-making. They might collectively send out a media release. The participants of the working group change and their roles are fluid and dynamic. We are used to having books published by such groups, in which either no particular authors are named or an alphabetical list of seemingly innumerable people appears without any clear idea of who did what or who did more or less – the latter all characteristics of more hierarchical organising.

In non-hierarchical structures conventional leaders are denigrated: there is no position for a domineering leader. Rather, leadership is a skill held by many, wielded temporarily and widely shared. The processes developed for assemblies and techniques for smaller meetings – such as hand signals and the authority of the collective – benefit from various trials of techniques of consensual and collaborative decision-making. Such techniques express the open, dynamic, solidarity-oriented cultures of transformative movements that include pacifist, environmental, social, cultural, ethnic, cooperative (housing and workplaces), anti-racial and gender-focused movements. Yet non-hierarchical ideals remain aspirational and challenging. Practice often falls short of the ideals of self-management, non-monetary forms of relating, social and environmental values, care and nurture.

In the past half-century, the divide between hierarchical and non-hierarchical forms of organising has lessened. A lot more philanthropic, not-for profit and profit-making firms, as well as governmental agencies, incorporate participative decision-making into their organisation. As simple techniques, such reforms can be deceptive and superficial in their ultimate effect on the ways governmental and capitalist institutions function. Yet the fact that they have spread is testimony to their influence, the influence of movements on societal structures, and the fact that horizontal and flat networks are neither unorganised nor disorganised but rather have their own formal typologies. This form of organisation is an ideal fit for spontaneous and anti-systemic movements made up of people from vastly different backgrounds and with different constraints on their time in movements that prioritise sharing knowledge and skills, and value flexibility and openness.

It is in the context of such political organising that degrowth has flowered as an open, decentralised and multidimensional network. Yet the nature of transformation demanded by degrowth has also, most significantly, spawned spheres of activity in individual, collective, resistance and big-picture, future-oriented action.

THE INDIVIDUAL SPHERE

We start with 'the sphere of the individual' although we might have started with any other sphere. Even if the sphere of individual activity is fundamental and representative, it really only makes sense when articulated (that is, placed in relationship with) the other three spheres. Any focus on the sphere of the individual only highlights necessary connections with collectives, with resistance and with an intelligible project – a shared, if fluid, degrowth imaginary. Instead, unfortunately, degrowth is often presented in a reductionist way focusing only on the individual sphere and, even more distressingly, by simply treating this sphere in a partial way. For these reasons alone, it is a useful place to start – to dispel common myths.

Voluntary simplicity

How can an individual embody and express degrowth? The essential answer is to adopt or adapt to a life of voluntary simplicity, that is, reducing one's environmental impact in an emancipatory and intentional way whenever, and however, possible. We are not all equal in terms of available choices. We live with constraints, sometimes due to relationships with other people and particularly due to economic factors, such as indebtedness. Pre-existing conditions can make it difficult or even impossible to individually take a radical step to live in voluntary simplicity, say refusing to use cars and relying on a bike or walking, or cutting down – even eliminating – use of a smartphone.

Because of the variety of circumstances different people find themselves within, the degrowth movement does not advocate a moralistic stance but rather a cautious open invitation to assess and renovate our lifestyles, when, if and as possible. Action in the individual sphere can consist of consuming less and working less yet enjoying a better quality of life. How might we achieve such an outcome? I might sell my car and, instead, use my bike, reduce my consumption of gadgets, clothes and only occasionally use devices and subscribe to associated services, minimise my consumption of meat, modify my travel habits (such as flying) and, ultimately, find myself less reliant on a monetary income.

At this point, I have created the opportunity to quit or reduce my working hours at my 'bullshit job'.[5] Thus I can re-appropriate free time for meaningful activities, such as caring more for others and looking after myself. I can contemplate, play games, hike, garden, read and engage in other such meaningful and pleasurable activities. We see here a dynamic situation in which a personal journey questioning one's needs mobilises the process of decolonising one's imaginary. Another world seems possible.

Identities and trends

Even so, changes in activities, thinking and consciousness in this individual sphere are not enough: we are not isolated individuals free to live our lives in whatever ways we want. We are embedded social animals facing contradictory circumstances and choices. We face social as well as material pressures, from mass media and social media – including advertising and cultural norms – regarding mores and habits. All of these pressures can make individual degrowth practices difficult to achieve. Changing our ways can create conflicts with relatives and friends or, conversely, offer opportunities to show others the constructive benefits of downsizing and changing pace. There are pros and cons, limits and potential, to all kinds of action in the individual sphere.

The emancipatory dimension of this sphere is influenced in structural and dialectical ways. This sphere highlights the significance of individual agency, pointing towards and integrated with other spheres. Even if the orthodoxy of a growth culture pushes us to desire, consume, produce and work more, there is a growing shared sphere of resistance. More and more surveys show rising expectations for quality of life rather than quantitative growth, and calls for more reuse, less work, more sharing and more recycling.[6] We observe, increasingly, public debates against planned obsolescence and aeroplane use. People are rallying in support of expanding train services instead of eliminating night trains, and various less frequented train lines, as has been common over the last few decades. There are strong bike rider movements in many cities, advocating for bike lanes and bike storage in place of car lanes and car parks. Vegan and vegetarian movements show a widespread preparedness to cut down – and cut out – animal flesh from our diet. It has even become *fashionable* to eat less meat and to prefer foods that are procured locally, in season and grown organically.

We can see, then, how individual action articulates with the cultural and social spheres to impact on norms and status. Contra Global North travel fashions and status symbols of the 2010s – to 'do' every country and see every city, to pick up memorabilia and to tell stories of such in cafés – as the decade closes more and more people are conspicuously deciding to slow down, take their bikes and discover their own locale and region. These are degrowth trends, whether the agents of such practices are aware of them as such or not. They illustrate people reconstituting themselves and their identities around practices compatible with degrowth.

Frugal abundance

This is the sphere of the degrowth individual who embraces, to some extent or another, 'voluntary simplicity' or 'happy sobriety' both based on 'frugal abundance'. Frugal abundance means letting go of work, consumption and environmentally unfriendly activities to make space and time to enjoy a rich quality of a life coextensive with a low ecological footprint. Frugal abundance is only achieved if we decide on meaningful limits and thus liberate ourselves from the frustrating 'always more'

Figure 3.1 Seven types of ordinary happiness
Cartoonist: Michael Leunig (Australia)

dynamics of growth cultures. Frugal abundance is intentional. We directly acknowledge Earth's limits by slowing down. We experience small as beautiful. We have re-appropriated our human well-being – allowing us, in the process, to become more humane (Figure 3.1).

The collective sphere

Most people active in the individual sphere also become active in the collective sphere, which consists of diverse local grassroots initiatives flourishing all around the world. In this sphere people re-appropriate, invent, adapt and experiment with new ways to produce, exchange and make decisions collectively. They act substantially outside the mainstream economy and polity. There are many inspiring and creative prefigurative 'concrete utopias' or, better, 'want-to-be-utopias', that is, utopias in the process of becoming, in different locations and sectors. Here we see distinct overlaps between degrowth activism and many other twenty-first-century movements with similar values and principles for living, such as slow movements, transition, Occupy, climate activists, the commons and land sovereignty movements.

Collective activities

The collective sphere represents an intriguing, multi-dimensional field in which degrowth individuals can develop skills and knowledge, create, invent, experiment, share, reconnect and care. Here the frugal abundance of a single individual can be multiplied many times over by an organised collective of degrowth-minded activists. Collective activities expand the opportunities, capabilities and capacities of degrowth households to enable collective sufficiency and collective governance, with greater benefits, economies and security. The individual goes beyond changing their, and their household's, practices to interconnect with others like them in collectives that encompass care, producing goods and services, and cultural activities. Here the necessary holism of degrowth is made apparent and its social dimensions are emphasised.

With respect to food, for instance, informal groups and formal associations have been forged to create food self-provisioning, from community supported agriculture to community gardens, from food cooperatives selling all kinds of organic food in bulk through to fermentation and food-preserving collectives, from local activities associated, say, with the Incredible Edible Network, to projects employing permaculture, organic, bio-organic, agroecology and agroforestry techniques.

As degrowth has become more widely known, many pre-existing groups pursuing such alternative techniques and approaches have adopted degrowth as a demand or to describe their activities. The book *Degrowth in Movement(s)*, an edited collection, traces the intricate connections with other such movements 'fighting for a good life for all beyond eco-modernism and growth, from a social-ecological and global justice perspective'; degrowth is at the core of this work because it 'symbolizes the most radical rejection of the eco-modernist and mainstream focus on growth, extractivism and industrialism'.[7] In the collective sphere these interconnections, co-influences, co-creations between and within like movements offers a rich environment for extensive experimentation with alternative techniques, relationships and ways of being.

Re-making the world and ourselves

The collective sphere features alternative activities based on goals of 'fair trade' and 'ethical production' from open and transparently run local production of high-quality organic goods to schemes involving non-speculative local and alternative currencies; from local exchange trading systems (LETS) to barter and 'no-exchange' (non-monetary) communities, from economies developed on the basis of reciprocity to sharing and gift economies, time-banks and solidarity economies. For such schemes to embody degrowth, they must go beyond offering a qualitative alternative to the mainstream to be more effective and efficient in genuine economies of elimination and reduction, both at source and within their mundane contexts. For instance, any car-sharing scheme that does not result in reduced travel and use of vehicles does not exemplify degrowth.[8]

The collective sphere encapsulates reusing and recycling with, for instance, various DIY and DIO makers' workshops and repair cafés. Such collectives focus on, say, bike building and repairs, or sewing second-hand materials using simple technologies and skilful hands and eyes. Repair cafés undercut trends to take away, throw away and planned obsolescence by offering repair services and, perhaps more significantly, training people to develop their own repairing skills. Such activities have sprung up around towns, sometimes as initiatives of the transition movement or of grassroots municipalism, that is, political associations based on direct democracy through neighbourhood assemblies and radical cooperativism.

Activities in the collective sphere include cooperatives upgrading dwellings with insulation and renewable energies. Ideally the latter are not only local but also low-tech, trialled as neighbourhood micro-grid systems rather than installed for individual households, and they apply solar thermal technology, which has environmental benefits over photovoltaic solar panels. Associated co-housing, alternative technology centres and other ethical, socially oriented hosting and hospitality projects are also types of collective-sphere activities. Many eco-collaborative housing projects feature intergenerational solidarity, say through mutual caring, hosting alternative learning workshops, engaging in sharing schemes and disseminating alternative ways of creating, living and exchanging.[9]

This is a far from exhaustive but, hopefully, representative sample of collective activities enacting degrowth values, strategies and principles. Last but not least are those collective actions that focus on transforming democratic processes. Such activities include groups that readily employ alternative participatory media, community radio and television stations, and digital interactive platforms for sharing information and co-creating documents and maps.[10] They experiment with community-based governance, say of commons, using consensual decision-making, 'sociocracy' (aka 'dynamic governance') methods and non-violent communication. Most collectives in various sectors and geographic regions adopt such horizontal and participatory forms of organising to some extent. Simultaneously, certain collectives are established primarily to develop, promote, disseminate and deliver programmes for learning participatory techniques of governance in all areas of life.

Networked collective activities

Significantly, especially in the past decade, many collective projects have expanded spatially through networking to develop more complex relations of production and exchange. They have created interactive networks of similar projects, and clustered dissimilar projects in complementary interdependent formations. In some locales in particular, such alternative collective enterprises are developing and interlinking in expansive and intensive ways, offering one another inputs and outputs using principles of fair trade, and production based on quality and care. Many draw on principles and processes associated with circular economies and social and solidarity economies.

Community supported agriculture, for instance, typically links farmers with food eaters living in a local urban centre through direct exchange networks. People sign up to purchase or pay in advance, or even purchase farm shares, in return benefiting from the farm harvest via regular deliveries or orders filled at pick-up locations. Within such farming communities, knowledge, skills, goods and services are shared across a range of agricultural pursuits. An abundance of raw produce might prompt a new collective based on processing and preserving, drawing on old and novel techniques. These expanded networks of collective activities are dynamic and fluid entities offering real-life experiments, experiences and images of the first stages of the degrowth project (elaborated on in chapter 5).

This collective level of activity is very dynamic, full of creativity and such initiatives attract rising interest. But not everyone can access, contribute to or benefit from such activities that tend to draw one into practices that conflict with the everyday demands of the dominant economic system. All members and participants face contradictions and make compromises with the capitalist market and state yet, wherever they exist, collective activities represent visible, and often persuasive, illustrations of degrowth in practice. They show what degrowth might look and feel like. They demonstrate solutions for more sustainable living, working and governing and that another world is possible even in very hostile contemporary environments.

THE SPHERE OF RESISTANCE

Even if individual and collective spheres are fundamental in establishing a degrowth society, there is another necessary characteristic of a degrowth transition – resistance. Only some people are in a personal and material position that allows them to make a move to a determinedly degrowth lifestyle. Even those who do practice degrowth are often keenly aware of concessions that they still make to the dominant system in their everyday struggle to survive. For instance, most of us are obliged to pay taxes to various governments and fees to service-providing agencies, even if we live marginally in relation to, and openly challenge, both conventional governments and markets.

We are expected to respect private property and contracts, private companies and public infrastructure whether or not they are fair or sustainable. Both as individuals and in collectives we find our practices at odds with mainstream market and state structures, challenging cultures and policies of growth. Our degrowth imaginaries urge us to resist the growth imaginaries, the growth forces, the growth processes and the growth institutions (such as banks), which all contribute to accelerating development, in short the destruction of our planet and social exploitation.

Resisting forces of growth, both growth imaginaries and the materialities of growth, degrowth activists have been particularly active in resisting the dominant form of transport in the Global North, private cars, and the plethora of massive road, motorway, flyover and parking infrastructures that they require. Degrowth Adbuster and anti-advertising campaigns signal resistance to the consumer society. Degrowth activists resist mega-infrastructure projects that either support the use of aeroplanes or expand commerce, such as new and expanding airports, new and expanding shopping malls. Degrowth advocates resist false solutions, such as so-called innovative 'smart', 'green' and 'sustainable' technologies, vocally and in print.

Significantly, the degrowth movement arose alongside the environmental justice movement, which has amalgamated concerns over the incidence of environmental damage and harms with matters of social disadvantage and inequity to find that poorer areas often suffer more deleterious environmental impacts than wealthy areas. The degrowth movement attracted increasing interest as a result of the deprivations and dislocations caused by the global financial crisis and ensuing austerity, recessions and depressions. First and foremost, within all degrowth reductions, is the principle of reducing inequalities of income and wealth between all of us as human beings. Resistance against inequity elicits further and more detailed concerns over corruption, tax evasion and the growth of public debts, which indebt entire nations of people.

Degrowth promotes deeper, inclusive and direct democracy. Degrowth activists resist violence, racism, xenophobia and sexism. In cooperation with other movements, degrowth activists develop non-violent strategies – such as direct actions, civil disobedience, blockades and flash mobs – to raise public awareness and to resist mega-infrastructure plans. Beyond employing classical forms of actions in street demonstrations, strikes and protest, degrowth activists engage in novel, customised actions (such as street theatre, and hiking banners on high buildings) in order to attract publicity – in order to alert people and raise consciousness, and implement constructive dialogue in public debates. The degrowth movement engages in these kinds of activities to *occupy* mainstream politicking and polity.

An example of resistance: Zone à Défendre

The sphere of degrowth resistance is neither wholly defensive nor wholly reactive. Degrowth activists do not act as a singular self-contained movement but merge with a collage of similarly associated, active and intentional change-makers. As such, degrowth principles and strategies of resistance have been demonstrated in an area known as the Zone à Défendre (ZAD) of Notre-Dame-des-Landes, a town close to Nantes (France). Here, several hundreds of activists occupied more than 4,000 acres of land, which was to be impacted by a planned airport. This internationally newsworthy blockade of long duration has incorporated a massive variety of social actors, many of whom are involved with, or aware of, the degrowth movement.

Surrounded by up to 4,000 soldiers at times, the ZAD activists not only defended the land but also started to experiment with and construct, in situ, an alternative utopia with direct democracy, local food production based on agroecology, local markets and alternative solidarities based on reciprocity and care, low-tech eco-constructions and energy production. These activities were intentionally emblematic of a commons, 'the common horizon that we share'.[11]

The main struggle of the ZAD has gone well beyond forcing the abandonment of the airport megaproject. Their struggle has been defined as 'not only a fight for the land, but above all for a way of living in common, which gives a whole new meaning to the idea of work or activity'.[12] The farmers, conservationists, unionists and environmental activists involved were keen to reframe public debate. Being against the airport could be construed as a selfish not-in-my-back-yard (aka 'NIMBY') protest. However, this coalition of forces was equally vocal against the worldview, the imaginary, behind such an airport project, its environmentally destructive construction and operation.

Active resistance was not only defensive but also constructively assertive. The protesters insisted on establishing a credible future:

> since the decision to abandon the airport was announced, the assembly hall is always full, filled by those who form the heart (not legal, but real) of the entity that we desire, and which will fight so that the forms of life that we have built here last and deepen. Forms of life that depend on ways of sharing that, to say the least, are not the norm.
>
> If there exists a place where the ownership of capital is not a source of pride and valorization, it is surely this zone. Many things are free here … [13]

The main challenge is finding strategies that enable such activists to co-construct conditions for an emancipatory, fruitful and calmly conducted debate at the same time as blocking further damage to planet and people. Indeed, as shown with the ZAD, degrowth activists try to infuse acts of resistance with clearer visions of their worldview and of the degrowth paradigm.

Those in the degrowth movement appreciate the complexity of challenges associated with socio-material change and use actions of resistance as opportunities to embrace more and more people in the co-construction of degrowth imaginaries, degrowth discourses and degrowth solutions.

The degrowth project: an articulated agenda

The degrowth movement and activists also inhabit a sphere of reflection and auto-critique. What, we ask ourselves constantly, is the agenda for degrowth? How can we define our 'project' to newcomers? Chapter 5 focuses specifically on the degrowth project; chapter 4 is on strategic matters, so here in chapter 3 it must suffice to address just one teasing question: how can we *achieve* democratic and peaceful transitions away from growth-addiction towards new degrowth models implementing degrowth principles?

This question is central to political debates and to discussions of degrowth with people in the street. It is central to academic, mainly transdisciplinary, theorisation and to empirical investigation ideally based on action research. Activities in all spheres contribute to co-developing a moral compass, social organisation and a contemporary voice for degrowth that, in turn, offers hope for and visions of a different future.

Many other questions beyond the central one arise from activities in each, and all, spheres. Is one particular strategy most appropriate and, if so, which? Which principles are key to driving or levering social, economic, institutional, cultural and psychological changes to enable the fruition of degrowth in practice? Discussions, debates and interventions around such questions occupy the sphere of the project and are central to creating a degrowth agenda.

As mentioned in chapter 2, degrowth initially became visible as a provocative slogan and quickly morphed into a matrix of increasingly well-articulated complementary thoughts driving action and seeking a real-world existence in the degrowth movement. Following similar dynamics and conditions for the different spheres of degrowth action, the sphere of the degrowth project or agenda can be readily defined but cannot be easily separated from the other spheres.

Individual action is inspired by the sense of a project, an endpoint. The collective sphere relies on individual skills and collective will but, equally, often evolves, and requires protection, from the sphere of resistance. Implemented by self-organising individuals, the collective sphere nurtures new forms of social agency and has a greater capacity to reach out to potential new supporters of degrowth. As demonstrated in the case of the ZAD, the collective sphere is protected by activities of resistance and resistance itself offers possibilities for expanding the notion and practice of degrowth.

It is argued that more clearly articulating the project of degrowth might enhance possibilities for consistent principles and characteristic processes to flower in various concrete degrowth experiments. Certainly this kind of coherence would improve opportunities to socially interconnect and materially integrate isolated and local initiatives into a clearer multi-functioning societal form. In developing a theory of change best suited to degrowth principles and conditions, the sphere of resistance offers rich cases to compare and contrast with the experience of the constructive sphere of pro-active collectivity.

In summary, the 'project' is not only speculative theory regarding what degrowth might look and feel like but also reflects on, and plays with, knowledge of the conditions, actions and results of concrete individual and collective experimentation. Here practice feeds theory just as theory, at another point, inspires action. A latent theory of change based on such praxis is gradually developing in the degrowth movement. Questions and claims around 'the project' drive radical debates defining paradigm transformation, ultimately evolving in multidimensional ways – intellectual, philosophical, economic, political, material, cultural and social.

INTERNATIONALISING DEGROWTH

The open degrowth network started and flourished in France. Step-by-step it became internationalised, primarily through activist and academic interest. Biennial international degrowth conferences, and a variety of other pop-up gatherings followed. Not surprisingly, the First International Degrowth Conference took place in Paris, in 2008. Since then, international conferences have been held in Barcelona (Spain) in 2010 and in Venice (Italy) in 2012, the same year that several universities in Montreal (Canada) sponsored the first international degrowth conference in the Americas. The Fourth International Degrowth Conference took place in 2014 in Leipzig (Germany) and, the fifth in 2016, in Budapest (Hungary; see Figure 3.2).[14]

These conferences followed degrowth principles in their organisation, planning and practice. They applied DIY, DIO and open source techniques. Each was unique, embedded well within local urban neighbourhoods and the natural local and regional environments to successfully symbolise the degrowth idea of 'open relocalisation'. Conscientiously

Figure 3.2 Fifth International Degrowth Conference banner
Graphic artist: Eszter Baranyai

co-organised by local people – most often a heterogeneous local collective of academics, activists and practitioners, including artists – the degrowth network Support Group worked to connect each and every dynamic local collective to the global degrowth movement.

Conference collectives go to great lengths to make sure that food is locally prepared from ingredients that are locally produced and/or obtained by dumpster diving. Processes are set up to seek and offer hosting by locals during the conference, which generally takes place over five days and in numerous locations. Local degrowth initiatives are promoted through tours and other types of events are informed and created by key degrowth activists. Sessions emphasise local political debates so that international degrowth communities can be inspired by, and contribute to, local experience and knowledge. At the same time, all international conferences emphasise strong planetary and international themes, which makes the gatherings genuinely 'glocal'.

Three large and complementary degrowth gatherings were organised in the second half of 2018. The Sixth International Degrowth Conference: Dialogues in Turbulent Times took place in Malmö (Sweden), 21–25 August. The First North–South Conference for Degrowth–Descrecimiento was held in México City

(Mexico), 3–7 September. A two-day conference in mid-September was held in, and with, the EU Parliament in Brussels (Belgium), with a preparatory one-day seminar of activists and scholars at the Université Libre de Bruxelles, and a wrap-up day at the European Trade Union Institute in Brussels, all stretching from 17 September to 20 September.[15]

The three parallel and complementary 2018 events show the influence of degrowth across several spaces. The biennial conference in Malmö focused degrowth within Scandinavia where, for example, the myth (and limits) of a green technological transition were of particular interest and, consequently, scrutinised. In Mexico, the organisers revived one of degrowth's fundamental pillars in the form of criticism of development to feature local indigenous movements fighting against extractivism, to produce on their own land in ways they chose, to enhance self-government and to celebrate the living idea of '*buen vivir*'.[16]

Last but not least, in 2018 the degrowth movement occupied one of the lobby-dominated sanctuaries of growth on the international stage, the EU Parliament, to explore what could – and could not – be done within this political space and its associated institutions. Frank exchanges ensued. Already, in discussions immediately prior to this conference, degrowth activists and scholars had become acutely aware of differences within and beyond the movement around questions such as: does postgrowth/degrowth mean post-capitalism? Such questions were raised again within the EU processes of engagement. Testimony to the success of the EU dialogue was an invitation to broaden and deepen such debates within this large institution.

The 2016 Budapest conference, with the motto 'Walking the Meaningful Great Transformations?', questioned degrowth in a post-socialist context, in a city amalgamated from two discrete cities around 150 years ago. Budapest is where East meets West to form something other than a pure mix of the two, a curious parallel with activists fighting for a vision of a new world while they live in a very different, and decidedly 'other', all too well-established, world.

Such conferences are organised in a spirit of openness, based on degrowth principles and radicalism, to constitute a living platform for dialogue.[17] Activists, academics, artists, politicians, inquisitive citizens and aspiring degrowth advocates meet to reflect, listen and learn about local and global realities. These conferences exemplify, experientially, the role of the degrowth network to facilitate better understandings of twenty-first-century environmental and social challenges. They embrace other similar movements and debate how to implement fruitful cooperation and to create useful synergies between distinct schools of thought in different regions, that is, distinctive socio-material conditions, all around the world.

Malmö was an ideal location for the 2018 conference due to its cultural diversity and social challenges associated with its industrial character. Immigration has meant that Arabic is second only to Swedish as a spoken language there. Those born outside Sweden make up 43% of the city's residents, migrants hailing from Africa and the

Balkans as well as the Middle East. Alternative, degrowth and grassroots activities are strong and varied.

In 2020, the Seventh International Degrowth Conference, to be jointly held with the 16th International Society of Ecological Economics Conference, had been planned for 1–5 September at the University of Manchester (UK). Due to the coronavirus, it was postponed to 2021. However, a complementary event – the Vienna Degrowth Conference, with the theme Strategies for a Social Ecological Transformation – did take place online in the 2020 European spring. Thus the crisis created an opportunity for the degrowth movement to experiment with techniques and technologies that avoid the environmental and financial costs of travel for international face-to-face conferences and have challenged the network to enhance an open relocalisation of their sphere of communication.

The intention of the Manchester 2021 conference is to invite participants to prolong debates that started in this very city in the Industrial Revolution around the contentious nature of capitalist practices, inspired by labour and cooperative movements. Here the legacies and failures of Marxian inspirations remind degrowth activists and scholars of the constant need to re-think their own founding principles, to be self-critical and to engage with and adapt to new developments.

DEGROWTH IN ACADEMIES

Degrowth became an active international field of research after the first conference in Paris in 2008. At that stage, the organisers of the conference had difficulties finding *any* peer-reviewed academic paper referring to degrowth! From this minuscule base, the number of journal articles, edited collections, sole and co-authored works has exploded, particularly in the last few years. Interest has been piqued and associations developed with scholars from several fields of research, such as ecological economics, sustainability studies, community and economic development studies, alternative economies, ecosocialist studies and heterodox economics.

In reference to the accumulative impact of this rise in academic contributions, Kallis et al. wrote in a 2018 *Annual Review of Environment and Resources* article that:

> Research on degrowth has reinvigorated the limits to growth debate with critical examination of the historical, cultural, social, and political forces that have made economic growth a dominant objective ... This dynamic and productive research agenda asks inconvenient questions that sustainability sciences can no longer afford to ignore.[18]

Indeed, in terms of research methods, degrowth is also forging a new path. Degrowth calls for transdisciplinary and interdisciplinary processes of investigation, and the constitution of anti-systemic logic. Degrowth practice has synergies with empathetic scholarly and participatory action research as well as critical governance and decolonial perspectives. Along with other edgy and provocative, if relatively marginal, fields of study, degrowth scholarship is dedicated to challenging and decolonising the comfortable imaginaries of certain well-established academic mind-sets.

As indicated in chapter 2, from Nicholas Georgescu-Roegen to Ivan Illich, from Jacques Grinevald to Serge Latouche, degrowth theorists have questioned science at its epistemological source. Can 'objectivity' be achieved? Is objectivity even necessary? Degrowth activist-scholars question the logic of market-based society's divisions of labour that have their parallels in disciplinary silos, many with reductionist perceptions of reality. Degrowth activist-scholars have sought research processes that exhibit a range of inclusionary methods, that involve the investigator in the investigation, methods such as participatory action research, citizen science and digital peer-to-peer sharing. All this in an effort to imbue degrowth studies with holistic approaches, to analyse and evaluate degrowth experimentation in the most appropriate ways. Here the 'activist-scholar' has been synonymous with degrowth research and theory.

'DEGROWTH IN MOVEMENT(S)': A PLATFORM FOR CONVERGENCE?

When degrowth first appeared, Paul Ariès spoke about it as a 'political UFO'. Indeed, it was misunderstood, rejected and demonised by green and left movements even if for different reasons and in distinctive ways. For the greens, degrowth was too anti-capitalist and anarchist. Why speak about capitalism or neoliberalism as the problem? Why not focus on the environment? For the more conventional union-aligned and hierarchically organised left, degrowth was suspiciously anti-work and anti-production. Why criticise industries which provide us with jobs? Why talk about bullshit jobs, part-time work and freedom from money when unemployment and precarity are both rising? Moreover, the old left sneer at 'unorganised' twenty-first-century movements such as degrowth – 'they won't amount to anything, they're not even organised!'

The practical left has a point. Unions of the twentieth century built their power base on struggles for full employment and maintaining or expanding their piece of the capitalist pie. That was the mainstream left project, the left agenda. But degrowth advocates have a point too: the traditional left has sat very vulnerably on a plinth of economic growth, compromising with capitalism to the extent of being co-opted, especially in adopting its economic language. Now that carbon emissions related to capitalist production threaten the future existence of humanity, the left and the greens – just like the EU Parliament in 2018 – are much more prepared to listen to degrowth theories and reconsider the meaning of degrowth practices.

In these left and green framings of new futures, anti-capitalism has greater currency and degrowth offers ways of both identifying and connecting the dots between the key material and social challenges of our century. Degrowth has a holistic view with a logical emphasis on key foci and steps to reach a pre-ordained end. Thus scholarly modes for degrowth researchers are transdisciplinary and interdisciplinary by the very nature of the case. When advocates of degrowth get into debates, they become acutely aware that dealing with our current challenges in singular and separated ways – or simply using economic frameworks – loses most of the key points and nuances of the radical and holistic degrowth perspective.

Too frequently there are mainstream media, political and economic discussions about accelerating climate change, biodiversity loss and the need to reduce consumption or waste without any reference at all to degrowth. Growth remains central in vain attempts towards decoupling, in creating so-called 'renewable' products, 'sustainable' services and, consequently, new 'green' jobs. The vast majority of familiar sustainability 'solutions', such as sharing and circular economies, are market-based – neglecting or distorting discussion of growth and degrowth. To attack growth is seen as commensurate with encouraging unemployment and deepening inequalities. Notice the cognitive dissonance when the logic of growth, which is coexistent with both the rise of contemporary forms of environmental unsustainability and the dynamics of establishing and maintaining inequality, is employed to solve these implicit consequences of growth! Anyway, the point is that degrowth has gone against the grain of established politics, left, right and green.

Confronted by environmental and economic crises, the human species also faces a well-acknowledged crisis of democracy and widespread incidence of anti-politics and populist parties of the right and left. Representative democracy offers individuals little power, but the chance to vote every few years for a 'representative' among a narrow range of options. Not only have political institutions, including political parties and the increasingly politicised media, been discredited – but also unions, progressive associations and non-government organisations have lost grassroots support to the extent that they have been corporatised.

Degrowth offers a radical invitation to examine the roots of our problems, a platform to connect the dots. The specific forms of inequality and unsustainability that we experience daily were bred within capitalism: productivism, consumerism, extractivism and materialism that have developed with added anti-social and anti-planet, racist and patriarchal, characteristics. Inequality and environmental unsustainability are not separate and cannot be solved separately. We are facing the collapse of a model of a thermo-industrial society based on more, always more.

As such, degrowth invites us to systemically analyse each of the crises we face, to broaden and deepen our understanding of economic and political structures, and to scrutinise the articulation of associated factors and forces. Degrowth prompts us to reflect and act accordingly. As neoliberalism privatises and eliminates public institutions, deregulates social welfare and retrenches or moves jobs offshore, we must develop more radical responses than the conservative (if legitimate) reaction of simply protecting our rights to public assistance, goods and services, and jobs.

In short, there is a need to go further than a defence mounted in traditional discourses and frameworks. For instance, if the plan is to relocate a massive automobile factory that has been the engine of the local economy, why not open up the local responsive debate? Why not decide on what constitute real basic needs for local people and devise ways of meeting such needs for all instead of making more cars for trade in precarious markets for unknown people in unknown places? Perhaps think more about the collective goods and services needed in this locale and how those goods and services might be produced say by a local cooperative? Degrowth treats news of collapsed firms as opportunities for conversations around new activities to create futures along sustainable and equitable principles.[19]

By way of an example, the union movement Earthworker Cooperative in Australia has been at the forefront of 'just transition' programmes for workers facing redundancy as non-renewable energy industries are replaced with renewable energy production and distribution.[20] Similarly, the yellow vest movement in France has argued that no green political agenda would be acceptable unless it included principles, practices and policies for social justice. Moreover, ecological taxes are only acceptable if they are embedded within a broader political project that addresses income inequalities and acknowledges that people on higher incomes have a greater capacity to pay such taxes.

Degrowth activists and scholars are particularly interested in 'praxis', in joining practice with theory. Degrowth has no fences and offers a platform for dialogue between numerous diverse movements. Degrowth acknowledges significant distinctions between the cultural and political contexts of each and every condition or struggle. Degrowth emphasises complementarities and distinctions between the different spheres of action – individual, collective, resistance and project-oriented. Furthermore, degrowth activists readily acknowledge a diversity of strategies appropriate within the glocal internationalisation of degrowth.

DEGROWTH IN PRACTICE

In summary, degrowth has evolved as a decentralised, multidimensional and open network facilitating emancipatory debates, dialogue and experimentation. Attempts at and arguments for a more conventional organisation of the degrowth movement have failed. Instead, the movement's radicalism has pivoted on an enduring provocative, interventionist, experiential and experimental stance. Combining various themes and arguments into a multidimensional demand, and surrounded by many movements with similar values, degrowth has long entertained a convergence, achieved by theoretically connecting the dots, and making practical alliances and identifying – if not forging – joint goals.

In the next chapter, we turn to some highly political questions and debates within the movement, namely: what political agenda, what political platform and which effective and desirable political strategies are most appropriate for degrowth?

CHAPTER 4

Political Strategies for Degrowth

In the early 2020s the international movement faces key political questions around the distinction of degrowth as a *discrete* movement, its political role and what that means in terms of political strategies for its future. Is it a problem that degrowth seems to have been more widely *influential* – rather than narrowly unique and crucial – within the plethora of movements and campaigns burgeoning this century? If so, can one conclude that degrowth's main role is simply to form a campaign, that is, to provide a solid critique of 'growthism' whether in capitalism or productivist market socialism? Alternatively, might degrowth strategically align with ecosocialist and post-capitalist visions, as a *leading* movement offering clear ways forward in our uncertain future?

In a generic sense, there are streams within degrowth with distinctive views on how the movement ought to organise internally and promote itself politically. Horizontalist, direct democracy and grassroots power-sharing streams are influenced by anarchism and gender politics. They conflict with more formal policy-oriented streams bent on proposals that might be considered seriously within contemporary political and policy discourses, by government bureaucrats and politicians. While some focus on novel disruptive models of marketplace enterprises, say, using digital peer-to-peer networking, others regard formal engagement and debate within mainstream and heterodox economics as crucial. Such distinctions become sharper when attempts are made to cohere around degrowth imaginaries, form degrowth agendas and take transformative steps forward.

How might the movement best articulate pragmatic strategies of 'doing', experimenting and prefiguring a degrowth society, engaging with parliamentary representation, collaborating with the union movement, and crucial non-violent direct resistance in terms of rapacious capitalist growth? Ultimately, matters of internal organisation and external relations are two sides of the same coin and essentially political. Deepening the analysis in chapter 3, this chapter on political philosophy and strategy lays the groundwork for debates on the degrowth agenda, the 'degrowth project', which is the focus of chapter 5.

ON INFLUENCE AND LEADERSHIP

In Lyon (France), three years after the launch of the slogan 'degrowth' and the enthusiastic campaigns around it, a general state of the movement (*Etats Généraux de la Décroissance*) meeting was organised in March 2005. More than 400 people joined the debate on how to organise the evolving degrowth movement. At this stage it seemed logical to follow the classic way, that is, organising a central structure, which would become both a platform and contact point for those within and outside degrowth. Yet this approach proved to be the first of a long series of failed attempts to formalise degrowth in conventional ways.

The degrowth movement is not readily organised using traditional forms. If degrowth has proved an explosive slogan in mainstream society, so too attempts to centralise its political structure have regularly led to fireworks, with horizontalism and autonomy winning out. This happened a few months after the general state of the movement meeting in France with the creation, and demise, of the Degrowth Party (already mentioned in chapter 3). The same stories can be told of developments in Italy and Belgium. Even if very influential, attracting large numbers of advocates and adherents, many practitioners, a lot of media and scholarly attention, degrowth has failed to materialise as a large mass movement in, of and for itself.

However, Paul Ariès has cautioned the degrowth movement about focusing on its own growth, saying it risks ending up like the frog in the Aesop fable that, in trying to prove it could be as big as an ox, burst.[1] In his popular talks Ariès often reminds the movement that, as long as degrowth ideas are spreading, then activists ought to be satisfied and simply proud of that influence, that they should not worry about not being able to point to a singular organisation and central headquarters for degrowth. In fact, as critics of the mainstream system and its dominating power structures, the degrowth movement needs to be careful not to reproduce the mistakes of capitalism's violent and arrogant domination, for instance by adopting its institutional forms. At the same time, creating some kind of appropriate and coherent form of organisation for the degrowth movement is necessary.

So, how might activists within the degrowth movement co-create such a structure, and shared processes that are genuinely effective and not counter-productive? How can the right balance be found between, on the one hand, being visible and impactful in the so-called real world yet, on the other hand, avoiding the risks of negatively centralising power within, and as, a movement? To be sure, the degrowth movement cannot solve contemporary political challenges using the same political tools and thinking that created the growth demise but, what are the alternatives?

The degrowth movement is not alone in facing these questions – they exist for all contemporary, emancipatory and emerging movements, from international anarchist movements to worldwide Occupy movements, from los indignados (M-15) in Spain to the French yellow vests (aka 'yellow jackets'), from the ZAD (Zone à Défendre) to Extinction Rebellion.

Change the world without taking power

After facing several tough organisational failures, the French degrowth movement consciously decided to 'own' its apparently disorganised nature and use it as a strong point for reflection. The movement focused on questioning a conventional sense of 'power' and how power manifests itself in various social and economic situations. Work on the concept and practice of 'autonomy' by authors such as Cornelius Castoriadis (1991) and John Holloway's *Change the World without Taking Power* (2010) has played an important role in degrowth activists' thinking.[2] In particular, Holloway's definitions of power as 'power over' and 'power to' encouraged those who experienced the state of the degrowth movement as a 'failure' to acknowledge the very real negatives of being successful in gaining power in conventional number-busting and leadership ways, that is, of wanting to have and exert power *over* others.

The vast majority of the degrowth movement feels freer and more confident in experimenting with a decentralised and horizontal network of small collectives and projects. These approaches are effective in creating *power to*, and avoid the destructive centralisation of power where all the movement's energy and effort is absorbed by gaining and maintaining *power over*. However, we still struggle, say, particularly with organising international degrowth conferences and associated media, to decide on common communication tools and questions around endorsing and formalising representatives for the movement. These challenges are always heightened in conventional everyday contexts where media, politicians and potential members of the movement ask us about contact points and sources for communications.

A DECENTRALISED HORIZONTAL NETWORK

The conscious decision to avoid conventional hierarchical power games must not mean abandoning power. The movement's perceptions and practices of power exist within a socio-political context that resists radical anti-systemic ideas such as degrowth. The movement exists to make a point and create socio-political change. Therefore, a balance has to be continuously negotiated between, on the one hand, the necessity to act within the system at the same time as, on the other hand, acting 'outside' and against the system.

Compromising too much ends in co-option, re-appropriation or falling into a plethora of other counter-productive traps. Yet being purist, fundamentalist or intransigent has dangers too. For instance, if considered solely from a mainstream perspective, it would be very convenient for capitalist forces to isolate degrowth into weak and small, apparently irrelevant, islands of radicalism. Therefore, the movement needs to engage in mainstream society in very strategic and self-aware ways. Correcting misconceptions of the movement need not be defensive acts but, rather, present opportunities for activists to explain the real meanings of degrowth and how degrowth is practised.

Fatal compromises have been made by other actors wanting greater influence. For instance, some large non-government organisations accepted funding from corporations only to find their activities compromised as a result. Moreover, certain green parties hell-bent on political power pushed representatives into government who, in fact, proved to be without sufficient skills, numbers or space to enact systemic change. Similarly, individualistic, narcissistic and purist actions – exhibited in certain types of survivalism where people focus on self-sufficiency and on building bunkers and accumulating stores to sit out an emergency – typically end up isolated. Such approaches oppose the degrowth spirit of mutual solidarity and support, of caring and collective sufficiency.

Ever since degrowth emerged, members of the movement have consciously walked a tightrope, sometimes engaging within the mainstream system, at other times existing outside it. Although not conventionally organised, the degrowth movement does have a clear organisational form as an *open*, *decentralised* and *horizontal network*. The movement aims to be free of the power-over strategies that conventional structures facilitate. On the one hand, members are offered experiential freedom to engage with and be accepted within the system so that they are listened to, have visibility and have a real-life influence. On the other hand, members are also free to engage in anti-power ways.

The movement maintains a culture that values freedom of speech and engagement in discourses over various differences. The movement finely balances being radical – through comprehensive root-and-branch anti-systemic critiques and experimenting with alternatives for broad-scale change – while avoiding extremism. Instead, the movement as a whole entertains and engages in both revolutionary and reformist tactics.

CULTURAL HEGEMONY AND CRITICAL MASS STRATEGY

Radical critiques of both failed revolutions and reformist strategies point out that it is not enough to simply seize and assume state power. The degrowth movement has integrated the Italian intellectual and politician Antonio Gramcsi's concept of 'cultural hegemony' into its strategic thinking.[3] Indeed, a main challenge for the degrowth movement is to radically influence society in cultural ways. Here the multi-pronged approach of applying the wisdom of that set of thoughts undergirding degrowth (chapter 2) through articulated spheres of action (chapter 3), plays a central role in decolonising the dominating capitalist imaginary – opening up people's consciousness to the possibility of creating an alternative world. Being active in the world makes degrowth real, enables fruitful conversations with observers and critics, and offers ways to integrate new members into the degrowth network (Figure 4.1).

The main goal of degrowth and related movements is to hypothesise, experiment with and co-create the conditions for change that ultimately leads to a critical mass support-

Figure 4.1 Activist Patrick Jones at a climate change rally, Melbourne 2019

Photographer: Meg Ulman (Australia)

ing postgrowth. The main goal is not to take power and exert *power over* because that runs a serious risk of the movement becoming a self-interested servant of its own power. The aim, instead, is to *influence*, engage with and incorporate, sufficient numbers of people to radically transform society. This is how the movement can change the world, neither by taking power nor by abandoning power to the reigning political and economic elites.

The authoritarian temptation

In analysing failures of utopian political thinking in the past, degrowth activists have identified a pitfall in the temptation to become authoritarian. Indeed, environmentalists occasionally express the fear that degrowth could develop into a type of eco-fascism. However, ever since it began, the degrowth movement has very clearly and openly distanced itself from such tendencies: 'Degrowth is democratic or it doesn't exist.' Experience has clearly shown that happiness cannot be imposed from the top down. Here the case of Thomas Sankara, who could be considered a pioneer of degrowth, is enlightening. In a popularly supported coup d'état in 1983, Sankara seized power over the French colonial Republic of Upper Volta state. He renamed his landlocked country in West Africa 'Burkina Faso', which translates as Land of Incorruptible People.

Sankara immediately implemented anti-capitalist, anti-imperialistic, ecological, feminist and emancipatory reforms, many in line with a degrowth political agenda. He forbade female genital mutilation, forced marriages and polygamy, appointed women to high governmental positions, and implemented educational and social programmes. Sankara supported agroecology techniques for reversing desertification, engaging in this effort French degrowth pioneer Pierre Rabhi. Rabhi later tried to run for the 2002 presidential election in France under the banner of degrowth. The country became self-sufficient in food within four years of Sankara assuming power. Lost in history is his inspiring speech to the Organisation of African Unity Summit in Addis Ababa in 1987 (a few months before his assassination) when he raged against the country's odious debt:

> Under its current form, i.e. imperialism-controlled, debt is a cleverly managed re-conquest of Africa, aiming at subjugating its growth and development through foreign rules. Thus, each one of us becomes the financial slave, which is to say a true slave.[4]

Sankara was, and still is, an inspiring visionary. But alone he was unable to culturally transform his society, as indicated in the following symbolic anecdote.

Sankara wanted his people to be autonomous and to feel proud of their culture, so he instructed public employees to wear traditional tunics woven from Burkinabé cotton and sewn by Burkinabé craftspeople. But, soon afterwards, he noticed that the public servants had not abided by his request. He ordered impromptu checks throughout his ministries and punished those who didn't wear traditional local clothing. In fact, under such duress, most staff members kept buying and wearing western clothes but each hid a traditional *faso dan fani* in their office to quickly put on, if they were subjected to a check.

One could make a class analysis here, or any number of other observations, but the salient fact is that the office workers had not culturally appropriated Sankara's great idea. Advanced in a hierarchical top-down way, his progressive reform failed. The authoritarian approach is a recipe for disaster. This is a clear case where an engaged, consciousness-raising and voluntary process driven by a decentralised, grassroots movement is likely to be more successful.

THE STRATEGY OF THE SNAIL

If a top-down approach is tempting, that is partly because we desire urgent, indeed immediate, change in order to face our massive climate, and biodiversity, emergency. This makes us prone to anxiously rush into transforming society as quickly as we can. However, the degrowth movement's response to the urgency of our conjunctural crises is to slow down, that is, 'to make haste slowly'. This strategy has evolved as such for two reasons.

First, even if it sounds counter-intuitive, we have learnt this approach from people dealing with catastrophes. Rescue workers and firefighters focus on a pre-prepared and methodical step-by-step approach so that they can glean all the significant relevant factors before acting appropriately in response to the emergency they face. These disaster professionals offer a model to the degrowth movement, which makes it less likely that we add to the chaos but rather, using intentional and deliberative processes, make haste slowly.

The main challenge for the degrowth movement is the temptation to follow the system's destructive and sterile rhythms. Instead, the response of the degrowth movement follows along the lines of the characters in the 1973 film *L'An 01 (The Year 01)*. The characters very consciously decided to come to a collective halt. Then, slowly, they reintegrate only those activities that prove absolutely necessary. Directed by Jacques Doillon, Alain Resnais and Jean Rouch, the film's message is 'Let's stop everything, and think about it: we're not sad!' (*'On arrête tout, on réfléchit et c'est pas triste.'*)

Therefore, in the degrowth movement's Platform for Convergence proposed by attendees of a gathering in 19 September 2009 (see Appendix 1), we find the following statement:

> In contrast with the classical strategy of taking power as a prerequisite to any change, we propose a radical and coherent idea: The Strategy of the Snail.
>
> ... the Strategy of the Snail implies that it is an illusion that acceding to power – whether in a reformative or revolutionary manner – is a prerequisite for changing the world. We do not want to 'seize power' but to act against the dominant structures and ideas by weakening their various powers and to create, without delay, conditions which will enable us to give full meaning to our lives.[5]

In this platform, the degrowth movement made a very clear statement of revolutionary transformation, so no-one need be anxious that a 'snail's pace' would mean any less radical change. An intention stated further down in the platform, which is reproduced in Appendix 1, is to make 'an immediate exit from capitalism'. Indeed, this aim leads on to our second reason for adopting a strategy of slowness.

In a modern society of spectacle, disruption and novelty – where there is massive social pressure to 'click', 'like' and 'retweet' – the strategy of stepping aside from immediate action and reaction definitely seems to risk the degrowth movement being written off. Nevertheless, for strategic reasons, degrowth activists believe that a carefully planned qualitative transformation will, ultimately, be the most effective way to successfully slow society down. The key dynamic behind growth is the money-making-more-money dynamic central to capitalist economies and societies. In this dynamic, where money seems intertwined with time, and time with money, there is an unstoppable urge to speed everything up. Thus, the decision to slow down is a qualitative counter-systemic strategy essential within the degrowth transitionary process. Slowing is, to some extent, qualitatively and quantitatively commensurate with degrowing.

Indeed, each of the strategies formed to progress degrowth must be read in multidimensional and contextual ways to gain a deep understanding of their authentic meaning. The strategy of slow conveys the message that, when faced with great challenges, being still and reflecting can save time otherwise wasted by acting impetuously on an inaccurate analysis leading to inappropriate action. At the same time, slow is clearly anti-systemic, which is sufficient reason for the strategy to be necessary and effective.

A STRATEGY OF NON-VIOLENCE

The question of whether to be, or not to be, violent has returned to the centre of current debates, given the frightening acceleration of climate change, and biodiversity loss, and the inability of current global capitalism to enact significant change. Questions around whether to act violently or not, and how to define precisely what acts are, or are not, 'violent', have been at the heart of debates on strategy by occupiers of the area planned for a mega-airport, la Zone à Défendre in the fields of Notre-Dame-des-Landes, north of Nantes (France), as described in chapter 3, a site of resistance for decades. Similarly, such questions arise constantly in movements like Extinction Rebellion, which is international, and with national uprisings such as the yellow vest movement in France.

The degrowth movement reflects on past actions and tries to learn lessons from cases regardless of their success or failure. The central questions raised with respect to violence often rotate around how best to be radical, how to avoid an extremist response and how to find the most effective strategies. A key risk of acting violently, while at the same time criticising the violence of the system that the movement opposes, is to find that public support quickly withers once the struggle turns violent.

Perhaps the right balance is found, instead, through determined, subtle and smart strategies? The degrowth movement is clear about rejecting any kind of violence, just as it is clear about rejecting any kind of authoritarianism. But here we need to define what we mean by 'violence'. Civil disobedience, direct action, sabotage, blockades, attacks against corporate symbolic property (say destroying a repulsive advertisement) are non-violent defences or mere provocations – if cleverly and collectively implemented – in order to open political debate, and as long as no-one is injured.

We have compassion for the despair protesters feel in a street demonstration that might compel them to attack the police or a bank as a veritable form of defence because the system of law and order and the financial system procedurally act as forces against them. Moreover, we simply don't live in the appropriate conditions for a dispassionate and solutions-oriented debate. Violence exists systemically, and systematically, within the capitalist societies in which we live. Historically violence has been an everyday tool of dominance, first to repress; second, to discredit movements; third, as a pretext to repress and isolate even more violently through injury, death and imprisonment. Thus, the circumstances in which people act, the societal powers people oppose, are already characterised by violence in terms of limitations imposed by police and police force and potential use of the military.

Within repressive environments such violence is readily accepted and the mere threat of violence is enough to discipline people not to challenge the status quo. Thus violence can emerge quickly, sometimes as the only form of defence. Even if an activist movement is not initially violent, if it is anti-capitalist, forces in the dominant system readily feel justified in provoking protesters associated with such movements. Therefore, the degrowth movement stands with a strategy of non-violence, also preferring non-violent communication strategies, so that we can genuinely claim that, if violence occurs, the dominant system will be the guilty party.

This question on degrowth positions with respect to violence and non-violence is framed by more general questions about how the movement can be radical, and tackle problems at their roots, without falling into counter-productive extremism.

TO BE RIGHT WHEN ONE IS ALONE, IS TO BE WRONG

More than ever, it seems, people exist within bubbles of 'truth', reinforced by social media provided 'free' to optimise advertising opportunities and money-making. Social media data and algorithms feed people's habits and preferences so that they see what they want to see and read news that contains messages that they are already likely to favour. People tend to develop a bubble of friends with whom they feel comfortable so that they are not challenged to grow, to be tolerant or to change. People existing in these conditions rarely meet others who disagree with them and thus easily dismiss them if they do. Degrowth advocates are challenged by this classic condition of alienation and respond with strategies of listening to, and engaging with, people whose imaginaries, ways of life and sets of values radically differ from their own. Such people are potential allies of degrowth. They help advocates understand better what capitalism is, and feels like, and help them to form and revise ideas about what degrowth might become. This strategy challenges everyone in the movement and prevents them from forming and living in a degrowth bubble of comforting ideas and people.

One might well ask whether this concept of people having a strong tendency to live in bubbles is really how they experience reality. Polls and voting associated with Brexit (Britain withdrawing from the EU), US President Donald Trump, the French right-wing populist and nationalist politician and lawyer Marine Le Pen, and Hungary's 'de facto supreme leader' Prime Minister Viktor Orban certainly suggest that this is the case. This state of affairs is also illustrated by the extent of mass media advertising and customised news and social networks both revealing and developing cultural and symbolic divisions in society. These online tools might be considered mass destruction weapons for democracy and dialogue, stretching across divisions that seem to become more entrenched along with rising economic gaps and deepening political divisions. Consequently, many feel more comfortable and compelled to travel by air to another part of the world than they do to get on a bike, or travel in a public bus, in their neighbourhoods where they live.

The degrowth movement needs to be skilful and find bridges to connect with people in cultures which may look and feel irreconcilable to degrowth advocates and activists but with whom we might find, or forge, a lot in common. Shared interests can often be found underneath the jargon used by both sides, and by discovering a common language to identify our commonalities. As examples of forging common ground, let's take the yellow vest movement and the question of bike use in Hungary.

Yellow vest movement

The yellow vest movement arose in October 2018, sparked by a government-planned so-called ecological tax on petrol, which threatened to raise fuel prices. The French state had underestimated the financial pressure that such a hike in fuel prices would put on those on a low income – especially those in rural areas – whose daily work and social life depends on using vehicles. Thus the yellow vest movement arose, organising numerous petitions circulated through mass social media, and started to block roundabouts all over France.

Apparently denouncing what was presented as a transitional reform to reduce greenhouse gases and oil dependency, the yellow vest movement was initially mocked, even demonised, by a large swathe of 'more educated' French, who typically lived in urban areas. But the protesters quickly fraternised and politicised their movement to avoid the trap set by Macron's government to push them toward the far right, where they would be contained and isolated. Throughout 2019, the yellow vests emerged as a movement for social and environmental justice, clearly illustrating the point already appreciated by those in the degrowth movement that there can't be an ecological transition without simultaneously improving social equity.[6]

From a rebellion against a discrete plan to raise taxes on petrol, proposals emerged from the yellow vest movement to prevent tax evasion by the rich, to set a maximum income, to tax aeroplanes and to redevelop regional train networks. As it happened, such top-down approaches from the yellow vests, that would so clearly rely on regulation and imposed constraints without either consultation or consent, were violently rejected within the broader movement, and were thus ineffective. But degrowth activists connected with the yellow vest movement to engage in dialogue and apply deliberative processes with more direct forms of democracy, respect and consideration. Advocating for these kinds of processes seems to have enhanced parallel political agendas with potentially much broader support and greater chances for success.

Contradictions in context: Hungarian bike cultures

We learnt similar lessons from the existence of dual bike cultures in Hungary, a country that boasts the third highest bike use in Europe after Denmark and the Netherlands. Hungary is unique because it has two radically different bike cultures: one urban and one rural; one chosen and enjoyed, one forced and painfully endured.

Critical Mass and later the I Bike BP movement both strengthened and made more visible the bike culture in Budapest and other large cities of Hungary. It has even become trendy to have a nice bike and ride up and down the city! People choose it, it improves their quality of life, health and happiness, and creates a sense of autonomy, pride and self-satisfaction. But, in rural areas where bikes are used more frequently than in cities, bike riding is commonly considered humiliating. You use a bike because you don't have enough money to buy a car and pay for the petrol that a car needs in order to be used.

Both urban and rural practices of bike riding align with an environmentally friendly political agenda for transport. But, the personal experience of bike riding can be very different, with enthusiasm and enjoyment of life at one end of the spectrum, and humiliation and frustration at the other end. When you consciously decide to get rid of your car and use a bike, you very rarely return to regular car use because your bike has become a very convivial tool in your life. On the contrary, when riding a bike is imposed as a form of austerity or thrift, as soon as you get the opportunity to get a car, you buy one. Advertising suggests that a car is more convenient and easy to use than is often the case and, humiliatingly, suggests that if you don't have a car, 'you are a loser without status'.

Here, again, we see the importance of constructing a positive and emancipatory narrative to contribute to a new cultural hegemony around convivial and freely chosen degrowth. Neither economic nor political force works. The degrowth movement accepts that matters of social justice need to go hand in hand with sustainability measures for one planet footprints to become attractive, and even feasible, for many people.

This is where the concept, practice and experience of 'open relocalisation' becomes relevant.

OPEN RELOCALISATION

Degrowth is about localising production and consumption, which often means 'relocalising' in practice. It is clearly desirable in ecological terms to minimise the transport of people, goods and services. It is totally absurd to have to travel to shop for goods and services that could be produced and consumed locally. It is equally wasteful to transport numerous products from a variety of places if they could be produced locally.

More significantly, relocalisation helps to break through some of the obscurity and illusions endemic to consuming goods and services that have been produced far away. When you go to a supermarket you often have to buy products without any clear understanding of how they were made. You do not know whether or not, or to what extent and in what ways, their production exploited ecological systems, say by using materials and non-renewable energy. The labels do not provide information about the conditions under which the people making them worked and were paid. This is not a moralistic statement but an alarming observation of the depth of secrecy in a private and competitive system of production for trade.

To relocalise production enables us to know how the things we consume have been sourced in terms of their materials, labour and energy use. Otherwise we are at risk of becoming a victim of being evil without directly doing evil – in the sense of Hannah Arendt's notion of the 'banality of evil' – of living in a false consciousness that all is well with the world.[7] The point is simple: Would anyone buy certain clothes, carpets or shoes if they had been able to observe the working conditions of the mere children who produced them? Would you consume a food product if you were aware of the specific mistreatment of animals that were involved with bringing it to your plate? Would you leave electric lights or appliances on longer than necessary if either the oil refinery or the rare earth mine for the so-called renewable electricity system that supplied your electricity was located in your neighbourhood or local region? Local production enables us to gain greater knowledge and real freedom to choose exactly what we buy.

In contrast to either top-down regulation or the free market perpetuation of unfair practices exploiting people and planet, a degrowth political strategy and vision focuses on radically transforming production, consumption and associated financial and commercial practices. In chapter 3 we discussed cooperative forms of working and communal distribution already being implemented by certain degrowth advocates, for instance through community supported agriculture and cooperatively organised housing, workshops and repair centres.[8] Such forms of self-governing production and distribution enable greater transparency and responsible decision-making over material inputs and waste, equipment and working conditions. Generalised, they would mean that we could make genuine choices to purchase ethically produced goods and services.

Yet degrowth-oriented regulations would be useful in certain circumstances. The power of advertising is widely accepted, for instance, bans of tobacco cigarette advertising have been introduced in many countries the world over. Regulating – even democratically elected bans on – advertising might be a very effective degrowth measure. The impacts of advertising are notoriously subtle and advertising contributes to cultures of growth by integrating a person's sense of identity with consumerism (Figure 4.2).

Figure 4.2 Advertising
Artist: Darren Cullen (UK)

Should we permit sport utility vehicles (SUVs) in cities where it is absurd, wasteful, unnecessary and environmentally destructive for a few tons of metal to use so much road and parking space just to move one or two humans (each around 65 kg) at an average speed that ends up comparable with cycling or travelling in a horse-drawn carriage?[9] Similarly, isn't it wasteful and inhumane to use petrol that is only supplied to us due to our governments having standing armies or supporting militaries where that oil has been sourced and refined? To curtail such practices, especially initially, top-down regulations could be used. Pete Seeger celebrated the City of Berkeley's Zero Waste Commission in his 'If it can't be reduced' song, where he went several steps further (see Figure 4.3).

However, taking a constructive approach, once degrowth was established, open relocalisation using degrowth forms of production would enable us to re-evaluate our basic needs, practices, production and consumption while

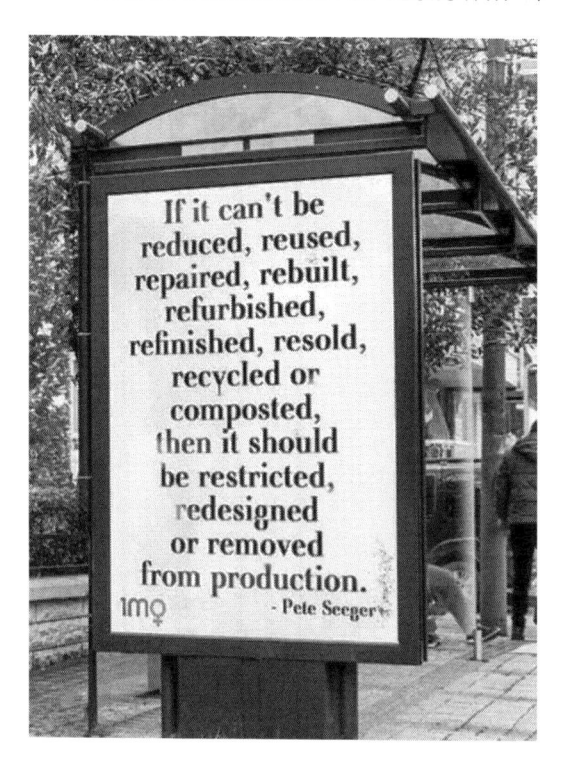

Figure 4.3 Pete Seeger – 'If it can't be reduced...'
Source: Twitter shared image, 2019

understanding the real human and environmental costs of purchases and facilitate us in consciously making enlightened choices. We will continue this discussion of open relocalisation in terms of practical proposals in chapter 5.

A SILENT TRANSFORMATION OF SOCIETY?

In February 2016, the left-wing French monthly *Politis* decided to produce a special edition assessing the impact of the degrowth movement. *Politis* interviewed most of the key French degrowth actors, asking them why such a relevant idea, that had attracted so much passion initially, appeared to have lost its appeal. The journalists focused on why degrowth had failed to maintain its originally spectacular visibility and to establish any clear and mainstream organisational form. However, as this investigative line raised further questions, a new narrative formed: perhaps what was really happening was an invisible underground revolution?

Ultimately, the *Politis* special edition appeared with the title 'Degrowth: A Silent Revolution'.[10] In short, the initial line of investigation was turned upside down or, better, right side up. The inquiry ended up revolving around the extent to which degrowth and associated movements had influenced political debates, so that even though 'degrowth' had not been absorbed as mainstream, it seemed that it was no longer a *provocative* idea. Perhaps degrowth did not need to become organised because its campaign was already obsolete?

Yet, if degrowth had become less controversial and more culturally tolerable, was it more politically acceptable? If degrowth was startling and attracted the limelight when it publicly emerged in the early 2000s, had the main point and ideas of the movement simply become clearer and begun to be seen as more rational over the last two decades, especially given the context of the global financial crisis, the increasing environmental crises, and debates on sustainability and climate change? Or, perhaps it was the strong critique of growth that degrowth advocates had managed to convey, or that was now easier to comprehend, rather than the constructive principles of a degrowth project, a degrowth future, that had become acceptable?

Although there is more than a grain of truth in the argument of a silent revolution, the argument needs to be muted because a genuine degrowth society is so far from being realised. Certainly, visible transformative steps have taken place in terms of consciousness, as confirmed by several public opinion polls. Surveys released in Europe in the autumn of 2019 showed that, with respect to utopias capable of addressing twenty-first-century challenges, it was the degrowth utopia that had most support from French people. First, in October 2019, *Odoxa* published a survey initiated by mainstream institutions – the British multinational insurance company Aviva, an economic weekly *Challenge* and the news media company BFMTV/Paris – to reveal that 54 per cent of French people surveyed supported degrowth, compared with just 45 per cent supporting 'green growth'.[11]

Second, in November 2019, the Observatoire Société et Consommation (aka 'Obsoco' or Society and Consumption Observatory) – with the support of the public environment and energy agency Agence de l'Environnement et de la Maîtrise de l'Énergie, the public investment bank Banque Publique d'Investissement (Bpifrance) and the E. Leclerc Chair in Future of Retail in Society 4.0 at the ESCP European Business School – invited those that they had surveyed to choose between three 'utopias'. In response, 55 per cent supported degrowth, 29 per cent preferred a security-styled utopia with top-down management of ecological impacts (in reality a type of eco-fascism) and just 16 per cent chose a techno-liberal utopia offering innovative transhumanism, green solutions and technologies within a liberal economy. In *Le Monde*, the Professor of Economy and co-founder of Obsoco Serge Moati concluded that, to a large extent, the idea of degrowth had been liberated from its negative images and associations, at least in France.[12]

Is there any evidence of similar changes of heart taking place in any other parts of the Western world exposed to degrowth ideas and practices? In 2016, Obsoco published the results of a similar survey that had been conducted in Italy, Spain and Germany, as well as France. Those results already showed such tendencies: 47 per cent for degrowth, 36 per cent for a 'collaborative economy' (based on digital platforms where users and providers engage in peer-to-peer transactions in activities such as car and accommodation sharing, time-banking and crowd funding) and only 17 per cent for transhumanism.[13]

Cultural changes are visible in trends associated with eating habits, such as eating less meat and eating more local, seasonal and organic food; avoiding flying; reusing, repairing and sharing; and becoming cyclists and walking more. We observe the success of books and films about local initiatives, such as the 2015 film *Demain* (*Tomorrow*), screened in French cinemas to more than 1 million people and translated into more than 20 languages.[14] This documentary shows how a new world driven by local citizen initiatives, like local currencies and alternative educational practices, is already under way.

These trends are related to other factors. First, even if denial persists at some level, there is much greater awareness of environmental challenges, especially climate change, loss of biodiversity and species extinction. The most recent reports from the Intergovernmental Panel on Climate Change and the Intergovernmental Science-Policy Platform on Biodiversity and Ecosystem Services (IPBES) are alarming enough, but more regular, extreme, intensive and prolonged climatic phenomena are making the crisis visible through events such as highly destructive and fatal bushfires and tornados.[15]

More debate on the limits of a neoliberal economy has arisen as a result of the 2007-8 global financial crisis and its consequences; the scandals around tax evasion revealed in the 2014 Luxembourg Leaks and the massive 2016 leak of documents known as the Panama Papers; the rising inequalities highlighted by Thomas Piketty in *Capital in the Twenty-First Century* (2017); and the social movements in Latin America, Europe, the Middle East and beyond, struggling for greater social justice. Last but not least, in the 2010s, we have observed an increasing number of radical critics developing novel concepts – already referred to in previous chapters – of the precariat and of bullshit jobs, concepts formed from unequivocal evidence of suffering due to abysmal or senseless working conditions.

A PEDAGOGY OF CATASTROPHES OR THE SHOCK DOCTRINE?

Naomi Klein's book *The Shock Doctrine: The Rise of Disaster Capitalism* (2007) describes how the neoliberal political agenda might deepen the catastrophes that our societies face by implementing even more exploitative reforms. Klein's 'shock doctrine' refers to situations where crises and catastrophes mainly bolster the neoliberal political agenda, thanks to its cultural hegemony and dominant media capacities. Recent examples include the phenomena of French president Emmanuel Macron (2017–), Marine Le Pen (president of the National Rally (2011–) and presidential candidate in both the 2012 and 2017 presidential elections), Boris Johnson and the 'success' of Brexit, and Donald Trump's headline-grabbing one-liners in his term in the White House, 2017–20. Yet French philosopher Jean-Pierre Dupuy's 2004 book *Pour un Catastrophisme Éclairé: Quand l'Impossible est Certain* (*For an Enlightened Catastrophism: When the Impossible is Uncertain*) speculates on how upcoming shocks and even catastrophes might offer 'opportunities' for social change agents to enact new realities.

Significantly, the terrifying increase of natural disasters of record proportions worldwide seems to have impacted more on popular debates on climate change and a preparedness to engage with serious imaginaries for our future than the increasingly detailed and alarming scientific works and recommendations that have appeared in the last decade. Observing the silent transformation under way, we can see these developments as offering pedagogical opportunities and possibilities for radical changes. Here, the strategic challenge for the degrowth movement is to become adroit at engaging in, and with, such cultural and political conditions – to develop a veritable pedagogy of constructive responses to catastrophe and to resist more 'shock therapy' from the state.

STRATEGIC READINESS: THE DEGROWTH AGENDA

With the undoubted rise in social consciousness, and with the severity of conditions that require action, writing in 2020 we find reason for hope. Unfortunately, at the same time, the global acceleration of both environmental destruction and socio-political and economic inequalities is depressing. As a planetary species, humans are still edging towards the precipice. The *Bulletin of the Atomic Scientists*' Doomsday Clock has just been set to 100 seconds to midnight.[16] Capitalism continues its violent exploitation of people and planet. We face the collapse of thermo-industrial 'civilisation'. Yet, attendant anxieties create more profitable avenues for manipulative and demagogic political agendas.

So, where to for degrowth? What items are on our political agenda? These questions, and more, are the focus of the following chapter. What does our ideal platform look like? What strategies do we need to address the main psychological and political barriers to adopting degrowth? As cultural support for degrowth grows, what are the necessary and optimal conditions to realise a degrowth model of post-capitalism?

CHAPTER 5

The Degrowth Project: A Work in Progress

Degrowth ideas and practices are shared and developed via a horizontal, multi-layered and open network of activists who experiment with degrowth ways of living and working, as well as advocating and campaigning for degrowth. Often stumbled upon by newcomers as a provocative slogan – a challenge to decolonise growth imaginaries – the concept of degrowth is based on a range of physical and social science theories that aim to address unsustainable and undesirable planetary, political and economic developments. In the first decades of the twenty-first century academic work, and even policy discussions, on degrowth have blossomed.

Consequently, the key challenge for this movement in the 2020s is to address the question: How might the numerous proposals made by various degrowth advocates consolidate into one cohesive degrowth agenda with clearly identifiable steps towards transformation? In short, what is the degrowth project? Given the diverse nature of the degrowth movement – and the complex variety of contexts within which activists are trying to bring about change – the answer is still up in the air, as is the case with any utopian quest. In short, the degrowth movement does not have a set holistic vision but, better, a constructively open approach to this question.

By way of comparison, even if aligned to degrowth in certain ways, the 'steady state economy' school puts forward a distinctly different and rather academic vision of a stationary economy substantially spelt out in quantitative economic modelling and regulated by the state.[1] In contrast, degrowth strategies and vision are more anti-economic, qualitatively focused and institutionally radical, as characterised by open relocalisation.[2] Therefore, the agenda for degrowth is open, and local autonomy allows for great variety within the application of degrowth principles of frugal abundance, respect for Earth's limits, and solidarity to meet the basic needs of all people, including beyond-material needs of communality, creativity and power. The vision, then, is the fulfilment of such criteria in a multiplicity of ways that has been referred to as 'pluriverse'.[3]

In this chapter we analyse some key methods for achieving degrowth that simultaneously create the necessary preconditions of radically reducing material and energy throughputs within over-producing and over-consuming societies. Representative initiatives discussed overlap and interact to create dynamic living systems of degrowth evolving locally, at least in embryo, as prefigurative formations. Preconditions include strong reductions in inequity glocally, frameworks for substantive democracy and the will to radically alter our economic processes.

This is a tall order at a point in history when powerful and inequitable capitalist economies and states dominate the planet. Despite slogans of freedom, opportunity and choice that are promoted as accompanying capitalism, these states are at most enjoyed by the few who are wealthy.

In contrast, the vast majority have little freedom because spare time is curtailed by working for sufficient income to meet basic needs; opportunities are limited given that unemployment and precarious employment are more frequent than secure work and incomes; others rely on welfare payments that severely constrain their opportunities; and producers decide what is on offer to consume, with low incomes restricting choices further (Figure 5.1).

Most significantly for the discussion here, living within capitalism one cannot choose a rational and appropriate future such as degrowth. The way forward for degrowth means running against the grain of capitalism, breaking down cultural barriers and encountering political, economic and physical resistance. The movement has to take account of these harsh contexts for change while trying to be constructive, confident and optimis-

Figure 5.1 Fuck work
Artist: Josh MacPhee

tic. Unsympathetic and oppositional contexts make the already political content of degrowth even more politicised and heighten the significance that the movement gives to 'autonomy'. The degrowth movement's use of the word 'autonomy' draws substantially from the work of Cornelius Castoriadis, along with Ivan Illich and André Gorz. Castoriadis was a twentieth-century Greek-French intellectual, an economist, social critic and psychoanalyst. As such, autonomy refers to collective self-organising governance with cultural relocalisation for the benefit of humans and ecosystems, and use of convivial tools creating post-industrial and post-development formations.

Thus, this chapter discusses key challenges in taking stock of the movement in 2020. We analyse a representative suite of proposals in terms of their fitness for purpose. We discuss challenges of a co-created agenda that entails reforming traditional structures and lively grassroots efforts creating direct democracy, needs-oriented economies and rich degrowth cultures. A cultural transformation, as in mass awareness, understanding and will, must be strong to support a non-violent deconstruction of current politico-economic conditions. So, we go on a journey where the project itself is a mirage, a reflection that appears when we are up close to any reflector but annoyingly disappears unless we position ourselves correctly and remain focused.

As for the degrowth project itself, the question for the movement is not so much how *far* we need to 'degrow' as how might we *best* 'degrow'. Just as significantly, the degrowth movement is committed to making experiences of degrowth desirable, comforting and emancipatory. To be inclusive and international are aims that go hand in hand. Arguably, global actions of degrowth that have focused on international conferences and movement alliances have developed more slowly than the multiplicity of local formations-in-the-making, which exhibit great diversity and common principles. Yet engagement between global movements with similar perspectives not only exists informally, among activists, but also has been a formal and productive goal.[4]

On the one hand, there are those who believe that 'the lack of monolithic definitions or unique policy pathways is a conscious choice of the degrowth community, which wishes to avoid the traps of the reductionism it confronts'.[5] On the other hand, the act of floating a degrowth proposal can attract support, direct trials and experiments, and guide ways forward. Proposals do not need to be emphatic, 'do-or-die' affairs. In this vein, a central proposal for an 'unconditional autonomy allowance' has emerged in France as a pathway, which includes a suite of measures on how to achieve a society in which we voluntarily and democratically decide on our basic needs and how to fulfil them sustainably.[6] We use this proposal to show how a series of actions might, in combination, gain momentum to create a degrowth future.

However, prior to outlining the unconditional autonomy allowance mobiliser, we summarise certain contextual challenges about agency for degrowth and a degrowth culture of inclusion at this particular point in history.

AGENCY FOR DEGROWTH: TAKING STOCK

An agenda for degrowth means taking stock of where and who we are, and what we might become. The first question, then, is one of agency. In the mainstream of the Global North, degrowth has emerged as an option for the future, even if unconsciously, that is, simply implicitly. As discussed in chapter 4, in terms of everyday practices and popular consciousness, decreasing use of resources and greater environmental awareness around waste and unnecessary consumption has grown. With the rising awareness of the ecological limits of the dominant growth-addicted system, people are voluntarily downsizing, decluttering, reusing, repurposing, repairing and sharing more. They are setting up an array of initiatives such as repair cafés and action groups to practically enhance sustainability by reusing materials and equipment, and minimising use of plastics or banning them. We observe ecologically friendly changes with people eating less meat and more locally grown, seasonal and organic produce, as well as changes in other consumption habits. Environmental policies and regulations related to producing common goods and services have been introduced that include sustainability criteria for housing. In terms of environmental sustainability achievements, local initiatives and non-violent direct resistance have been impressive, especially set against international governmental inaction on climate change.

Indeed, climate action, namely reducing carbon emissions is a remarkable example of the kind of change that the degrowth movement is all about.[7] The movement for climate justice offers degrowth activists a great opportunity for strong interventions with the potential to catalyse greater acceptance of degrowth ideas and practices. The climate justice movement calls on governments and citizens to take care that measures to cut carbon emissions do not impinge on the livelihoods of those on lower incomes; people with low levels of consumption are the least responsible for the rise in the level of carbon emissions in the atmosphere. The climate justice movement emphasises the disproportionate responsibility for carbon emitting activities in the Global North while many people and lands in the Global South are more vulnerable to the impacts of climate change. This approach follows principles developed by the more long-standing and holistic degrowth movement, offering an opportunity for advocates and activists to promote degrowth.

A 2020 *Guardian* survey of respondents to the simplistic question: 'Climate, inequality, hunger: which global problems would you fix first?' had a minuscule 3 per cent voting for work and economic growth – in the last place – while the majority (54.1 per cent) gave their first choice to 'saving the planet'. The second choice was for equality (14.5 per cent) and 'end poverty and hunger' (13.3 per cent) a close third. This internationally accessible online survey ran from mid-January with tally results as at 22 February.[8] If this survey is any indication, the vast majority of people favour degrowth concerns with the environment, equity and satisfaction of basic needs – the exact reverse of economic and political elites' preoccupations with the economy and jobs at the expense of ecological considerations. Other international polls show that environmental issues are gaining greater traction; a 2019 *Guardian* poll conducted in eight European and North American countries showed that the climate crisis was of greater concern than migration or terrorism.[9]

Similarly, French polls cited in the last chapter showed a strong swing explicitly in favour of degrowth. While degrowth is not yet the subject of other national polls, a 2015 survey of readers of the reputable From Poverty to Power Oxfam (UK) blog showed similar levels of support: 80 per cent regarded degrowth 'a good idea' even if 38 per cent did not like the word itself.[10] Last but not least, in January 2020, the Edelman Trust Barometer report found that 'more than half of respondents [56 per cent] globally believe that capitalism in its current form is now doing more harm than good in the world' with an 'alarming' growth in distrust within 'the mass population'.[11]

Moreover, governments adopting sustainability policies throughout all sectors and regions have inspired, and been encouraged by, cultural and practical changes showing greater respect for the environment and concern to act in more ecologically friendly ways. State action includes advice to householders on reducing consumption and waste, environmental footprint assessments, and applying targets for reduction. Limits and rationing, such as temporary or permanent water-saving regulations, are being accepted as rational in many regions. Undoubtedly, frustration with current impasses in terms of mainstream responses to our environmental crises prompted the Post-Growth Conference at the European Commission in September 2018, an indication of influential people wanting to learn about degrowth.[12]

Conversely, the drivers of growth and commensurate degradation of Earth continue apace as states perversely promote growth, with targets to increase GDP and 'sustainable development' rather than adopting 'sustainable degrowth' and 'post-development' perspectives.[13] Thus, the radical political, economic and cultural transformations necessary seem like pipe dreams in the face of a very global capitalist mega-machine imposing its destructive agenda. Consequently, it seems to many in the degrowth movement that there is a massive gap between aspirations for radical and desired change and our strategic capacity to achieve change. Institutions, groups and people aiming for change tend to feel weak, isolated and depressed, as encapsulated in the mid-1990s line that it has become easier to imagine the end of the world than the end of capitalism.[14]

Beyond cultural barriers, such as the advertising that infiltrates our media and surrounds us in urban built environments, and the denigration of environmental advocates as 'trouble-makers', there are real, material and social inequalities – a power imbalance that seems to reduce anti-capitalists' capacities for attracting support and enacting change. So the degrowth movement is challenged to create a vision of alternatives that a growing number of people might find both sensible and attractive enough to appropriate and drive. The degrowth movement needs to effectively deconstruct the social and material barriers to degrowth messages and activities that accumulate in everyday life like a blanket of alienation. How can degrowth advocates and activists successfully puncture all those psychological, symbolic, cultural, economic and social barriers that create the daily dissonance within which most people feel trapped?

Capitalism has always created scarcity and inequities, through private and exclusive enclosure. The degrowth movement stands for the opposite: equality and openness, revealing genuine individuality. We want to create radical and frugal abundance through sharing while maintaining ecological sustainability that is an insurance for such abundance in the future.[15] It stands to reason, then, that the agenda of the degrowth movement must be welcoming. Beyond being provocative and visionary from an idealistic standpoint, the degrowth project needs to be clear, simple, subtle and practical. The movement favours voluntary measures and action, deconstructing and reconstructing out-of-date institutions without force but, rather, with non-violent action. Above all we need to work for inclusion.

INCLUSION

Classifications of degrowth advocates and activists as 'primarily middle-class, relatively highly educated young whites' might well be correct but there is little hard evidence one way or the other.[16] Even, if correct, such characteristics need to be rationalised and put into context. Activists can almost always be charged with being non-inclusive by the very nature of the case. If you are a total victim of the system, you will be trapped by socio-cultural and economic circumstances that give you no time, energy or even inclination to explore beyond survival within the current system. If you do make a great effort to join the movement, you might not feel comfortable with other activists who seem relatively privileged in terms of time and money, and you might feel resentful that you're risking a lot more than your fellow protesters to devote time to degrowth campaigning and other activities. If you have ventured this far, you might well find it is too wearing and dispiriting to keep participating.

In a degrowth transitionary initiative or prefigurative formation, say a food cooperative or a community supported agriculture collective, participants can become disappointed that the project's membership is not representative of the whole spectrum of socio-cultural diversity. It is always a challenge, especially within a deeply anti-systemic movement such as degrowth, to have wide representation if only because the movement does not attract entrenched agents of the system, whether they be passive victims or self-interested agents. At the same time, one finds that many degrowth activists do not fit the caricatures bandied about by journalists, politicians and critics. These include caricatures of wealthy, urban and well-connected graduates; layabout, unemployed, so-called professional protesters; and, even worse, 'environmental terrorists'. For those willing to speak out about their daily problems, degrowth cultures engage in collaborative listening, problem-solving and caring to support other activists in their everyday challenges. Mutual aid abounds.

Embedded in the movement's attention to reducing inequality is acknowledgement of greater misery in the Global South and a need to reduce inequalities between the Global North and Global South. Often this is expressed as a call for Global North shrinkage, not only to respect Earth's limits but also to allow for all people in the Global South to meet their needs. Moreover, the degrowth movement strongly supports and calls for the self-determination of all peoples. Decreasing consumption in the North has been challenged because perceptions of a 'North versus South dichotomy is counterproductive insofar as it glosses over the fact that a substantial part of the elites and the growing middle classes in the South live a Western, growth-oriented mode of living'.[17] Furthermore, by what mechanism can decreasing consumption in the North translate to improving the conditions of those in need in the South? Centuries of capitalist imperialism have effectively frozen skewed trading relations, imposed exploitative relations on Earth and a multiplicity of peoples, imprinting inequity writ large.[18] For several centuries growthism has been a major force worldwide depriving Indigenous peoples of their lands and cultures.[19]

From another vantage point, many peoples in the Global South already practise degrowth, both consciously and unconsciously. Indeed, degrowth advocates of both the Global South and disadvantaged regions of Eastern Europe have had some similar experiences and, equally, face some similar challenges.[20] Chapters in collections on housing and food for degrowth show how citizens of Vanuatu unconsciously practise degrowth by building simple and appropriate structures collectively; how architects in South India run counter-growth, to build using degrowth criteria associated with tools, work and materials; how Kenyan women in cities maintain traditional food choices, preparing and sharing household knowledge skills and tasks to counter strong commercial influences; and how food self-provisioning in Central Europe expresses degrowth values and relationships.[21] Global North activists are learning from Global South degrowth activists, incorporating their wisdom, engaging and sharing models for moving forward. The First North–South Conference on Degrowth was held on 4–6 September 2018 in Mexico City with the motto 'Decolonizing the social imaginary'.[22]

Degrowth activists are frequently well-meaning people from various backgrounds and many pay a price for their activism, which they feel acutely in their daily lives. 'Coming out' as a proponent of degrowth often separates them from certain people while, at the same time, enabling and emboldening them to act against the great challenges of inequality and unsustainability. If the degrowth movement stands for decreasing inequalities at the very base of the distinctions between Global North and Global South, then being an active member and advocate of an anti-systemic movement such as degrowth can be argued to be a *declassing* experience. The elite and managers of capitalism spurn those who would challenge the system they command as traitors to their class and bureaucratically organised unions and workers feel uncomfortable with the idea of working less and in alternative ways, for both experience and tradition tell them that they must fight to be workers, to remain employed, to get a full-time wage and a bigger slice of the pie. Being a degrowth activist sets one apart from traditional class identities as the movement fights for a class-free world.

The degrowth movement exists in a broader context of much more subtle and critical discourses of class identity than a middle-class sandwiched between a proletariat and ruling class. The big brush '99%' represents a plethora of identities and struggles, with precarity a characteristic of rising numbers of income earners (e.g. in Australia 1 in

4 workers is a casual).[23] This so-called precariat is a heterogeneity of the aspirational, the dispossessed, would-be salaried workers, radical anti-capitalists and conservatives. They are a cluster of marginalised people who only share insecurity, and experience it variously. The degrowth movement of protest and action is replete with radically disruptive analyses of class politics that go well beyond bureaucratic unions and parties yet selectively collaborates with sympathetic unionists and politicians in radical networking and actions.

Agendas of inclusion and radical transformation set the degrowth movement within current activism that seeks – with certain difficulties and self-criticism but also great cohesion – to combine a rich diversity of multiple struggles in a future where all humans can flourish diversely. This polity punctuates and confuses class analyses. Prominent activist-scholar Angela Davis has usefully cast the vision of such transformative movements as 'antiracist, anti-capitalist, feminist, and abolitionist' giving 'priority' to grassroots movements 'that acknowledge the intersectionality of current issues – movements that are sufficiently open to allowing for the future emergence of issues, ideas, and movements'.[24] Thus, the 2016 international degrowth conference in Budapest prompted the 2020 publication of a 'dictionary of social movements and alternatives for a future beyond economic growth, capitalism, and domination' in which degrowth appears 'in Movement(s)'.[25]

Many current activists are students who have a fluid and uncertain class identity. Sometimes, irrespective of their class of origin, they are automatically considered middle class based on the assumption that they will qualify to fill professional positions. However, having a doctorate no longer means immediate employment as it did in former generations. Sometimes students are marginalised within analyses as neither reflecting their class background nor the class position that they are expected to occupy – written off as neither one thing nor another. Moreover, in the 2010s, as in the late 1960s, students have periodically risen to shake political elites with internationally newsworthy calls for greater democracy, direct representation and independence.

The degrowth movement represents such uneasily categorised 'change-makers', incidentally illustrated in this comment on student degrowth activists by degrowth academic Giorgos Kallis:

> There is a vibrant community and this is an irreversible fact. In Barcelona 20–30 of us meet frequently to read and discuss degrowth, cook and drink, go to forests and to protests. We disagree in almost everything other than that degrowth brings us together. In the fourth international conference in Leipzig, there were 3500 participants. Most of them were students. After the closing plenary, they took to the shopping streets with a music band, raised placards against consumerism and blocked a coal factory. Young people from all over the world want to study degrowth in Barcelona. If you experience this incredible energy, you find that degrowth is a beautiful word.[26]

Even if systemic and cultural reasons exist for the degrowth movement not being or seeming inclusive, it is still a challenge that those within the degrowth movement want to overcome, if only to develop an agenda that incorporates everyone and addresses their basic needs. This is where a platform mobilised by an unconditional autonomy allowance (outlined in more detail later in this chapter) and sharing work and care, can demonstrate ways in which the degrowth movement could apply measures of equity, whether through state reforms or local degrowth group activities in a vast variety of places. A tension between strategies of state pressure versus grassroots action remains in play, but together they create a non-violent scissor movement for deconstruction without force. Thus the degrowth platform is realistic and has the central goals of reducing inequality as well as reducing production and consumption. Moreover, the degrowth culture assumes that the democratic skills of listening, engaging with and re-evaluating any such platform ensue.

This discussion on inclusion shows why our project is, and must remain, a work-in-progress. Being inclusionary necessitates being open to adopting widely supported enhancements, and even reframings, of our agenda. The movement's attention to reducing inequality and Global North advocates incorporating Global South activists' concerns means a permanently adaptable approach. Simultaneously, the degrowth agenda must highlight key principles and be framed clearly, to illustrate its 'alternative' culture embracing diversity, sharing economic and social burdens, and operating through self-governing processes and institutions that preserve autonomy yet operate globally, in other words the degrowth movement strives to be 'glocally' embedded.

REPOLITICISING SOCIETY AND RESOCIALISING POLITICS

Prevailing political and economic elites in growth-oriented economies have been seen to reinforce inequalities and govern and manage their affairs with little transparency. There has been a rise in that type of populism that sees the masses seemingly voting against their interests, as in the 2019 election in the UK won by the Brexit-trumpeting Boris Johnson and the supposedly protectionist Donald Trump who became the US president early in 2017. Such bombastic leaders who aren't scared of 'saying what they think' typically promise a return to ruling the world in their external negotiations, harking back to the imperial grandeur of a simpler age of national pride and economic security, decorated with racist overtones and closed border policies. Such leaders break with the two oppositional party forms typical of representative democracies of the twentieth century by flouting party discipline and policy platforms and, instead, seeking cross-class support and personally making policy decisions on the run.

Similarly, the French yellow vest movement illustrates a rising divide that is 'beyond class' in that it cannot be easily explained in a class analysis. The main drivers in identifying as a yellow vest are subjective defiance toward 'the system', the elite, and belonging within a more broadly and contemporarily defined 'other' than class (interpreted as reflecting personal economic interests).[27] Neoliberalism has strengthened individualism and economic competition. 'Fraternisation', a key word of the yellow vest movement, rapidly politicised those who had been depoliticised by neoliberalism. The yellow vest movement began with a narrow campaign rejecting a top-down undemocratic, humiliating and fake carbon tax. Yet the yellow vests ended up debating policies to set a maximum income and blockading hubs of the US multinational Amazon, one of the global 'Big Four' technology companies.

The yellow vests and unions were originally suspicious of one another but later joined forces in strategic and respectful ways.[28] The main learning, for the degrowth movement, from the yellow vest and other populist movements, is to recognise a clear need to create trust, informal solidarities and to repoliticise society while resocialising politics. This includes overcoming that state of meaninglessness identified in Castoriadis' concept of 'insignificance' (and its threat of incipient barbarism) by facilitating cultural and political re-empowerment.[29]

Within the current conjuncture, the degrowth movement has made interventions in, and alliances with, compatible political and environmental movements of the twenty-first century and degrowth activists have developed formations at the level of the collective sphere (chapter 3). However, at least from the point of view of certain French degrowth advocates, greater support might well rely on an approach in which an unconditional autonomy allowance acts as a mobiliser, a transitional path, with some preconditions already under way dependent on where you live.

AN UNCONDITIONAL AUTONOMY ALLOWANCE

The unconditional autonomy allowance refers to a monetary income and/or in-kind right for all, from birth to death, to ensure a decent and modest, frugal, way of life. Moreover, the unconditional autonomy allowance constitutes a mobiliser in the transition toward sustainable and desirable models of degrowth societies and must be complemented with local deliberation over production and distribution.

This approach incorporates a suite of schemes such as an unconditional basic income, an acceptable maximum income, work-sharing and free access to basic services. Associated changes involving regulation include a sharp reduction in advertising sales of goods and services, and staged reductions in uses of resources such as energy and water. Radical regulation of advertising – representing more than US$560 billion in 2019, with North America the largest market – would move from initially heavily taxing advertising to, ultimately, banning it.[30] This measure is advocated by the UK Special Patrol Group (Figure 5.2).

Such policies, regulations and structures assume a central state or well-established level of governance based on subsidiarity, prioritising power at a local level. Indeed, proponents of an unconditional autonomy allowance approach sensibly argue for working with reforming current structures to initiate and facilitate a degrowth transition. For instance, more substantively democratic, participatory and direct forms of governance – techniques often referred to as 'deliberative democracy' – are assumed to evolve as organisational supports. In this, proponents take a primarily institutionalist approach, given that the Regulation School has influenced the French degrowth movement.[31] It is also a proposal meant, like the

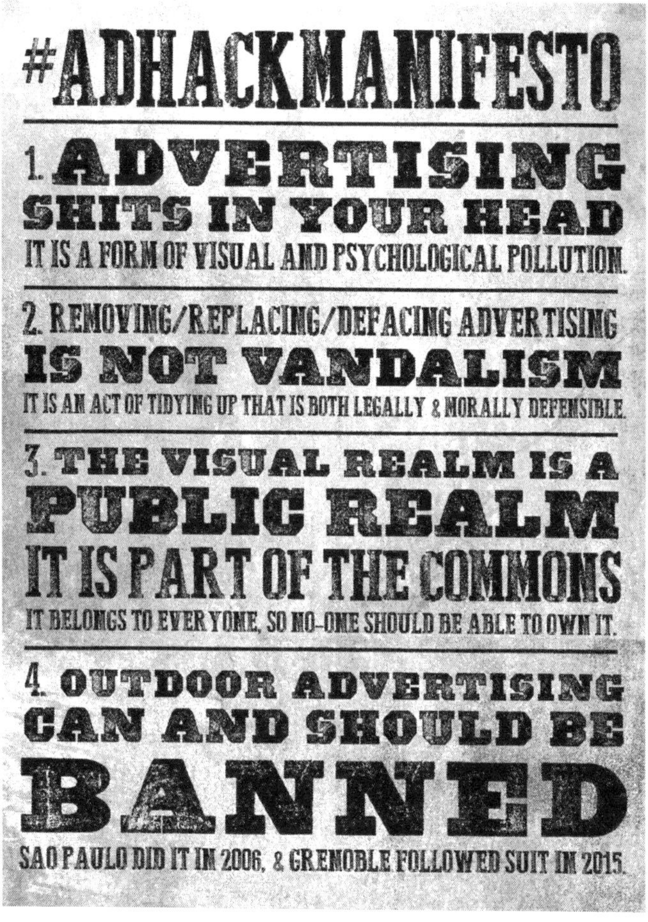

#ADHACKMANIFESTO

1. ADVERTISING SHITS IN YOUR HEAD
IT IS A FORM OF VISUAL AND PSYCHOLOGICAL POLLUTION.

2. REMOVING/REPLACING/DEFACING ADVERTISING IS NOT VANDALISM
IT IS AN ACT OF TIDYING UP THAT IS BOTH LEGALLY & MORALLY DEFENSIBLE.

3. THE VISUAL REALM IS A PUBLIC REALM
IT IS PART OF THE COMMONS
IT BELONGS TO EVERYONE, SO NO-ONE SHOULD BE ABLE TO OWN IT.

4. OUTDOOR ADVERTISING CAN AND SHOULD BE BANNED
SAO PAULO DID IT IN 2006, & GRENOBLE FOLLOWED SUIT IN 2015.

Figure 5.2 AdHack Manifesto
Source: Special Patrol Group (UK)

word 'degrowth' itself, to serve the purpose of encouraging discussion and debate, 'flying a kite'.

Perhaps the most significant aspect of this unconditional autonomy allowance proposal is that it is meant to assume, as well as inspire, grassroots activities such as agricultural and non-agricultural productive cooperatives that are oriented to satisfying local basic needs, that is, substantially self-provisioning cooperatives; local production and repair of simple, appropriate and convivial tools and goods in maker cooperatives; and community-based exchange systems, involving local currencies, sharing and gifting. As in theories and practices of social and solidarity economies, many of these economic ideas of degrowth proponents draw on theories of Karl Polanyi that aim to 're-embed' the economy via twenty-first-century movement practices of active democracy and the degrowth concept of open relocalisation.[32]

OPEN RELOCALISATION

Open relocalisation places humans and ecosystems at the centre, as substitutes for where monetary value and capitalist enterprises stand within capitalism. Human needs and ecological limits are key determinants of degrowth practices, which revolve around human relations with one another and with nature, and where coexistence is central. People's needs encompass autonomy, power, and transparent and respectful collaboration in co-producing and sharing what is produced so as to fulfil basic material needs and beyond-material needs. The degrowth movement openly accepts people's responsibility for one another and for nature more generally. Advocates and activists acknowledge a need to live as modestly as possible, and see co-production and sharing as ways to live more efficiently and effectively, that is, more securely.

The degrowth movement's current organisation persists as a horizontal, multi-layered and open network of activists. Consistent with this form, many degrowth advocates envisage a future based on governance at the most local, immediate and decentralised level possible, an organisational principle of 'subsidiarity' where considerable power exists at the grassroots level. In this model, production and distribution of basic needs is highly collective and localised around sufficiency. This concept of grassroots governance complements a transformation of gender relations based on sharing responsibilities and expectations of 'care' not only in social spheres but also in environmental, ecological, spheres. Healing Earth, regenerating nature and facilitating ecological well-being is most easily, effectively and efficiently conducted at a discrete local level.

Localised self-provisioning has environmental and social benefits – ecological efficiencies, community-based autonomy, cultural richness, self-governance, caring for one another and celebrating in frugal abundance. Processes of direct democracy are implicit in the concept of an unconditional autonomy allowance because deliberations on what 'basic needs' might be need to be specified locally, that is, within the context of how they can be met – even if basic needs can be defined by certain universal generalities. A 'commoning' approach means deciding how to collectively manage local means of production, from land and tools to the effort expended on productive activities. All of these elements cluster together in the holistic notion of 'open relocalisation'.

Significantly, certain degrowth grassroots activists are explicitly anti-state, or non-state-oriented, having encountered too many statist barriers. They believe ardently in a grassroots revolution that might effectively take over all the remaining necessary functions of the state by supplanting its rationale for power locally and establish the existing horizontal, open and multi-layered network on a firm institutional base.

Consequently, the following pieces of the unconditional autonomy allowance jigsaw puzzle can only sketchily show how a variety of concrete schemes might come together. These measures rely on, and would drive, a cultural transformation, perpetually decolonising growth imaginaries and propelling institutional, economic and political transformation. They are seen as ways to achieve an emancipatory, democratic and peaceful transition towards societies that are ecologically sustainable, desirable, relocalised but connected, open, convivial and autonomous. Key questions remain across every formation and over time: What basic needs do we need to produce? How do we produce them with optimal ecological and social benefits?

There is no magic recipe but a large range of pathways within which a universal autonomy allowance offers a coherent convergence of complementary levels and approaches.

AN UNCONDITIONAL BASIC INCOME

A key barrier to both self-empowerment and collective empowerment arises from the demands of our economic system, that is, the necessity of working to gain our basic needs and to repay debts that arise because average incomes are not sufficient to cover the costs of education and further training, buying a car or owning a dwelling. For most, the risks of losing a job or being demoted mean a rat-race world, marginalising personal involvement in political action. For repoliticisation, degrowth must show how people's essential needs can be met in alternative ways, that is, that forced employment is not necessary to fulfil the basic conditions for a dignified life. Solidarity can deconstruct alienation, offer hope and develop trust. To open discussions and debates in this direction, we propose an unconditional basic income.[33]

An unconditional basic income would offer every individual enough in money or in-kind goods and services, for a decent life from birth to death. The idea is not new. More than two hundred years ago, the English-born Thomas Paine – who spent many mature years in the US and some years in France during the era of the 1789 French Revolution – propagated the idea of a basic income as restitution for loss associated with the institution of private property. He argued that a minimum income was necessary for any citizen to accomplish the republican principles of freedom, equality and fraternity.[34] Likewise the proposal here links the rights of citizens to political autonomy and influence, with the necessary fulfilment of their basic needs. The provision of an unconditional basic income is an act of solidarity, a refusal to abandon anybody, and assures the dignity of everyone.

This idea recently re-entered political debates and has been experimented with in a variety of places.[35] It has many supporters, particularly within younger generations, as offering an opportunity to avoid forced employment in a system that they reject and as security to launch meaningful activities, such as establishing a degrowth future. Like any measure, an unconditional basic income can be distorted and co-opted, for example, the measure proposed by Milton Friedman was really a 'negative income tax', which aimed to support capitalism and make it more efficient.[36]

The unconditional basic income is distinct from social security measures such as unemployment, sickness or disability benefit, or a retirement pension, because there is no eligibility requirement to be actively looking for work, to be ill, to have a disability or to be retired. Moreover, even if an unconditional basic income is effective and coherent in terms of a degrowth agenda, it is only a mobiliser within the degrowth project, which requires other measures to be applied in unison, specifically an acceptable maximum income and unconditional basic services.[37]

ACCEPTABLE MAXIMUM INCOME

In as much as paid work exists, and currently working for money is the primary way many people satisfy their basic needs, an acceptable maximum income is a degrowth demand.[38] This is more than a cap on wages and salaries but would extend to income from other sources, including inheritance, with grandfathering or similar staggered application. In the degrowth movement a 1:4 difference between minimum and maximum incomes has certain support. This would sharply reduce current inequalities. Research from 2017 by Credit Suisse highlighted that the world's richest 1 per cent own 50 per cent of wealth, and in the UK the wealthiest 1 per cent own around 25 per cent of UK wealth in contrast to the poorest 50 per cent with less than 5 per cent.[39]

The degrowth movement advocates fairer distribution as more effective than redistribution even if redistribution remains a useful tool for reducing inequalities. Of course, initial measures in mainstream economic contexts are specific to monetary relations typical of capitalism. Beyond monetary measures, the movement proposes techniques common in highly developed welfare states, emergencies and socialist countries – access to a set amount of goods and services per capita either at a set price or gratis.

UNCONDITIONAL BASIC SERVICES

Unless an unconditional basic income is supplemented with special benefits for those with special needs, for example, associated with ill health, then health and associated support services need to be free.[40] In other words, unconditional basic services must be part of the degrowth package. They would include products such as water, energy supplies and access to decent housing, health, education and funeral services. Degrowth advocates acknowledge a need to de-commodify what is too important to be administered by an irrational market. After all, why do we pay the same price for a litre of water for drinking, cooking, food gardening and washing ourselves, clothes or dishes as for a litre of water to wash a private car, fill a private swimming pool or water a golf course? Shouldn't we reserve certain sectors to supply basic needs to all?

If this principle seems simple, questions around how to achieve the free supply of those goods and services necessary for dignity abound. In terms of quantity, there would be a basic per capita maximum allowance, depending on the region. For instance, given that, for climatic and other reasons, some regions have a lot more water available to them than others, there would be greater per capita allowances in areas with fewer ecological limits or future sustainability risks. Democratic re-appropriation of water management might initially proceed via a progressive pricing system based on principles of free access for a reasonable usage and extra costs for extra usage or misuse, measures that would be applied using meters. Similarly, ultimately, energy supplies would focus exclusively on local renewable sources, such as solar thermal, relatively low-tech and crafted for territorial energy sovereignty.

Following the principle of diversity, the degrowth movement has a liberal approach to the design and implementation of an unconditional autonomy allowance. It might be delivered by a variety of institutions, in-kind, through use of formal or alternative monies. As such provincial, local or even state governments, regional degrowth formations or commoning might organise production and distribution based on a right to access maximum amounts.[41] Those self-governing commons producing for collective needs would decide on the definition and constitution of 'basic needs', not a simple material assessment but, following the kinds of criteria established by scholars such as Manfred Max-Neef, incorporating non-material basic needs comprising criteria such as 'dignity'.[42]

BASIC NEEDS

Already basic need provisioning is evolving in situ and being debated. The right to land and housing has been highlighted by political squatters who occupy buildings to create social spaces as well as accommodation, where feasible incorporating productive activities based on methods such as agro-ecology and permaculture activities to fulfil basic needs.[43] Re-appropriation of land can occur through state requisition. Christiania in Copenhagen developed a commoning system for housing, later modified towards a more mainstream model as a condition of the settlement's 'normalisation', that is, legalisation, by the state. This case highlights struggles between state and activists seeking autonomy and freedom from market forces.[44]

More conventional paths to create housing pointing in the direction of a degrowth future include eco-collaborative housing models such as community land trusts, cooperative housing, cohousing and ecovillages strongly oriented to ecological and social justice.[45] In Zurich, where cooperative housing has an impressive twentieth-century history and accounts for more than one quarter of apartments, the New Cooperatives movement is specifically oriented to sustainable futures, with careful consideration given to building and living sustainably.[46] An extract from a radical proposal for future consumption appears in Box 5.1, demonstrating how distinct a sustainable city or rural town might look like in a degrowth driven future. Eco-collaborative housing models share an orientation away from the monetary values and relationships of markets towards land and housing allocated through systems of self-governance based on use rights and social and material production to fulfil basic needs.

Housing with sufficient land or a community-based enterprise, such as community supported agriculture, both contribute to provisioning of relocalised, seasonal,

---

Box 5.1 A lifestyle menu

For reasons of ecology and justice, a typical Swiss lifestyle menu might look like this:

- 20m^2 of private living space
- 2.5m^2 of communal space (space shared with others so, e.g. 20 sharing = 50m^2)
- no cars
- no flights
- 6 km train travel per capita daily, within Switzerland
- a boat voyage of 1000 km yearly
- 15 kg meat yearly
- 20 L milk yearly
- 70 L water daily
- 3 hours internet weekly
- 1 printed newspaper daily shared between 50 people

The different factors are partly interchangeable: eat less meat, but enjoy a car trip, reduce your living space for a short-distance air flight and so on. On the whole these limits call for a completely different lifestyle, which requires a different residential, territorial and institutional setting, that is, a postgrowth setting.

(And, of course, there is no cap on sharing, reusing, creativity, friendship, love, care and so on.)

Source: Slightly adapted from New Alliance. *A proposal.* 2019. New Alliance site – https://newalliance.earth/a_proposal.pdf)

organic and sustainable food production based on agro-ecology, organic and permaculture principles and highly plant-based diets. Complementing such food and housing initiatives, maker workshops and repair hubs focus on making and using appropriate, convivial tools, applying open-source knowledge and creating low-tech, hand-made or home-made furniture, utensils and clothes, operating on the basis of sharing and recycling (see Box 5.2). As indicated in the text of both Box 5.1 and 5.2, walking, cycling, scooting and a modest amount of train travel are favoured forms, which necessitate extensive changes in infrastructure and settlement planning. All such initiatives arise in contexts where an onerous system of monetised production and exchange turns citizens away from mainstream trade and financial arrangements towards local community-based currencies or similarly alternative exchange systems.

ALTERNATIVE FINANCIAL SYSTEM

In 2017, global debt amounted to US$184 trillion (nominal terms), that is 225 per cent of GDP, greater than US$86,000 per capita or 2.5 times the average per capita income. The richest countries were the most indebted, with the US, China, and Japan responsible for more than half of global debt, more than either of their proportionate shares of global output and double their combined contribution to world population (24 per cent). Meanwhile the International Monetary Fund confirms that private sector debt, which has trebled since 1950, is 'the driving force

Box 5.2 Degrowth formation: Cargonomia

Cargonomia evolved when three pre-existing social and environmental enterprises in Budapest and its surrounds joined together as a degrowth formation-in-progress:

- Cyclonomia DIY Bicycle Social Cooperative
- Zsamboki Biokert, an organic vegetable farm and sustainable agriculture community education centre distributing vegetable boxes weekly to Budapest food communities
- Kantaa, a self-organised bike messenger and delivery company

This formation is a self-organising focal point for individuals and communities interested and active in producing sustainable food, promoting and using low-carbon transport, bikes and a bike culture. Operating on sustainable, convivial and fair trade principles, food is distributed via direct marketing and by cargo bikes that are made by Cargonomia. Members can borrow such cargo bikes.

Cargonomia offers an open space for community activities directed towards sustainable transitions, conviviality and degrowth, hosting DIY and self-sufficiency building workshops, discussions and cultural events. Cargonomia produces research, for instance on care work, and members were key to organising the Fifth International Degrowth Conference in 2016.

The formation incorporates an open localisation approach, welcomes initiatives that will develop local production of basic needs, and aims to inspire and facilitate empowerment with concrete alternatives to standard profit-driven social and economic systems. In all these ways Cargonomia is developing towards the realisation of a degrowth future using an approach of open relocalisation.

Source. Adapted from data on Cargonomia website: http://cargonomia.hu/?lang=en

behind global debt', with China in the lead.[47] Capitalist growth and monetary debt are tightly integrated.

The 2008 global financial crisis resulted in recessions, depressions and austerity. For instance, structural adjustments were imposed on Greece in a very violent and spectacular way, driven by the Troika, the International Monetary Fund, the European Central Bank and the European Commission. In the name of public debt, social programmes, health and education budgets were cut with severe impacts on Greek citizens. Movements such as the Committee for the Abolition of Illegitimate Debt ('CADTM') that focuses on Global South debts, have been calling for public debt audits for decades.[48]

The full dimensions of debt include exploitation and extermination, involving slavery, war (as a business with associated industries) and the genocide of indigenous peoples. Degrowth proposals include establishing transparent audits of private and public debt, and moratoriums as necessary. Foreign debt has been an imperialistic tool to further economic interests. Especially in the case of default, terms and conditions of foreign lending or credit renewals have led to interference in domestic policies and to the imposition of capitalist political agendas, which have recently resulted in the de-development of European countries, such as Greece.

Money as a token of credit and implied debt is fundamental to capitalist economies where money 'saved' institutionally, typically in legal banks, is in fact lent out to gain interest. Moreover, banks create money through lending activities. In a competitive system of private firms producing for trade, bank loans and the interest they demand drive economic growth. Furthermore, the financial system and financialisation rampant this century means systemic pressure towards speculative bubbles, necessarily bigger – and with broader impacts – the more that the capitalist system expands.

The debts of capitalism include literally hundreds of billions worth of tax evasion hidden under the euphemism of 'optimisation'. Moreover there are incredible levels of inequalities between large corporations profiting from the system and the smaller, local businesses that tend to contribute to society more fairly and appropriately. Beyond the inscrutable practices of big business are the hidden debts of the market economy – human and ecological debts resulting from exploitation and appropriation of limited social and environmental resources, destabilising ecological cycles and leading to extremely dangerous 'tipping points'.[49]

Driven by the necessity to instil a sense of responsibility for planetary and local limits, degrowth advocates ultimately argue for a radical takeover of the economic system. The degrowth movement envisages a new model of society based on principles other than growth and greed, a society which supports social and environmental justice and facilitates both redistribution and a desirable democratic transition. The movement seeks alternative governance of the financial and banking systems, aiming towards their democratic regulation or re-appropriation. Degrowth advocates and activists call for rethinking money creation, its governance and the role of the reserve bank. In terms of reforming the current system, one proposal is for full-reserve banking, zero or even negative interest for money creation for emancipation from a debt culture, from growth-driving dynamics and the associated devastating impacts on environments. Thus, the degrowth movement proposes community-governed and community-oriented banks in preference to current monetary and financial institutions.[50]

ALTERNATIVE FORMS OF EXCHANGE

Those experimenting with creating fairer economic models have adopted a range of alternative forms of exchange considered more complementary to their activities and more expressive of their principles than formal currencies, which are generally issued or regulated with respect to their issue by state central banks. Alternative exchange systems allow credit for work and goods, arranged between members of a formal organisation. Alternative exchanges include community-operated non-speculative local currencies such as local exchange trading systems (LETS), based on a local unit of credit; community currencies, such as the Bristol Pound or Totnes Pound (2007–19); and time-banking based on a labour-hour as the unit of credit, such as Time Credits (UK). Time-banks allow multilateral exchanges of work using the measure of per-hour work, say one hour of plumbing, language tuition or childcare.

Schemes involving local exchange systems and currencies are long-standing and widely practised outside (as well as within) the degrowth movement, but often represent only a low proportion of the total exchanges or provisioning for each member. LETS involve multilateral exchanges with a central accounting system. Members are encouraged to be relaxed about being indebted to the system because all members are either in credit or in debt constantly. Frequently, a small amount of internal credit is withdrawn to remunerate administrators or associated costs. Currently, most alternative exchange schemes run in parallel, even in competition, with legal currencies – and have created headaches for taxation departments, especially when their members only agreed to pay sales taxes in their own currency! Degrowth activists have adopted and adapted a variety of differently organised and managed alternative currency schemes as appropriate for a degrowth transition.

More radical and comprehensive financial and exchange models created by activists from the degrowth movement and beyond include the Cooperativa Integral Catalana, which has evolved in the Barcelona region with their 'ecoxarxes', exchange nodes, operating complementary currencies.[51] Informal economies of solidarity and reciprocity seem peculiarly appropriate for degrowth economies where production focuses on provision of basic needs in local economies. Others in the degrowth movement theorise over non-monetary exchange systems, including gift economies, where there is no form of monetary accounting at all – proposing wholly non-monetary economies based on direct democracy.[52] Informal and lightly organised sharing schemes can be as simple as creating a space in an apartment or neighbourhood block where people leave what they don't want for collection by someone else who can use it. At the other end of the spectrum of sharing models are fully collectivised communes with 'one purse' and high levels of self-provisioning.

For the purposes of degrowth, ideally the alternative community-based currency cannot be subject to the act of saving. It might have a use-by (expiry) date or automatically devalue over time, finally to become worthless. Ideally, it cannot be lent, especially as in the sense of invested. The primary driver of capitalist debt is investment borrowing. Thus, the currency system is designed to avoid speculation and simply promote exchange and circulation. In fact, any investment becomes the prerogative of the community.

Indeed, currently degrowth activist campaigns protest against investments in advancing Frankenstein-type dystopian transhumanism, so-called artificial intelligence and growth-oriented mega-infrastructure. They call for a moratorium on all such investments and, wherever possible, implementation of public deliberation on where to invest time, energy and resources in future. The degrowth movement calls for transparency in establishing key criteria of what constitute unbearable environmental and social impacts project by project. The movement argues for open and democratic determination of priorities and preferences for projects, and the upfront identification of those individuals and communities that projects would benefit (and disadvantage).

The unconditional autonomy allowance approach would integrate a complementary local currency based on a degrowth charter in order to promote and support local production, services and exchange. The charter would be universal in its general principles while each locality would have a unique currency. The goal would be to reintroduce close communal cooperation within regions to create more direct democracy and sustainability, reconnecting people, and co-creating trust and feelings of security. Strong relationships of obligation and solidarity, and principles of unconditional support for all to maintain a minimum level of dignity constitute the foundation for both participative democracy and a degrowth transition in which we all identify our basic needs and how to satisfy them in fair and sustainable ways.

CONCLUSION

In *The Great Transformation* (1944), economic anthropologist Karl Polanyi theorises that within capitalism the market economy has become the dominant ruling force. He argues that, in order to recover genuinely democratic political power and to manage economies in a more environmentally responsible way, we need to 're-embed' the economy through the state exerting authority over the market. Driven by a need to decolonise our imaginaries, and to disarticulate capitalism – from 'the top down' with state measures and 'from below' with autonomous grassroots activities – the degrowth movement aims to re-embed the economy through democratic deliberation on our basic needs and how to fulfil them in fair, sustainable and convivial ways.

Degrowth and aligned movements offer a platform for dialogue, convergence within spheres of action (chapter 3) and appropriate strategies within which the unconditional autonomy allowance is but one proposal. The unconditional autonomy allowance blends an unconditional basic income, a maximum acceptable income and unconditional basic services with a set of radical reforms enabling us to re-embed the economy under direct democratic control along with social relationships of obligation and care, sharing, reciprocity and localised economies. These types of measures could evolve, and in certain places are evolving, always customised and adapted to local, regional, cultural and political specificities. In summary, the degrowth project envisaged in terms of an unconditional autonomy allowance covers various basic needs, as outlined in Appendix 2. The unconditional autonomy allowance mobiliser incorporates numerous measures, such as a maximum acceptable income, and could be implemented as a transitionary process that takes place step by step, as shown in Appendix 3.

As such the unconditional autonomy allowance proposal here does not stand for 'the degrowth project', but rather offers a sense and certain logic to proposals emerging in and around degrowth. In its soul and design, a degrowth project articulated around an unconditional autonomy allowance could be implemented in decentralised and relocalised ways, involving a revolutionary reformist approach based on dialogue, deep listening, consensual decision-making and appropriate action. It would be based on the logic of open relocalisation and how to change the world without either taking power or abandoning power. It would require cultural transformation and grassroots activities both of which are already under way, but all this demands structural, economic and political transformations yet to be born.

Postface:
'Now Is the Time of Monsters'

The US political intellectual Immanuel Wallerstein wrote a memorable analysis of contemporary 'anti-systemic movements' in 2002, the very year that degrowth became visible in France as '*décroissance*', after its decades-long gestation. Author of the classic four-volume work *The Modern World System* (1974–2011), Wallerstein iterated that the global structural crisis had prompted 'an "age of transition" – a period of bifurcation and chaos'. In short, many typical challenges and, therefore, characteristics of twenty-first-century movements of resistance and change could, and should, be contrasted with movements of liberation of the nineteenth and twentieth centuries. We had moved into a distinctly new era: 'one of deep uncertainty, in which it is impossible to know what the outcome will be'.[1]

Writing in mid-April 2020, it seems like the collapse of Western civilization is in full swing, recent events exposing the very cleavages degrowth and similar movements have recognised and highlighted for decades. The spread of the coronavirus exposes the risks and vulnerabilities of advanced capitalism – an open and global market, profit-oriented strategies, financialisation associated with economies of scale and social dumping. Now we have seen in sharp relief how, if China's economy slows down, the global economy is shaken to its core. China supplies many key products for the external market – including active ingredients for medicines, and facemasks – and so its production, and any hiccup in it, impacts on all kinds of international sectors. As in a row of dominoes, as each and every country – Italy, the US, Spain, France and the UK – was affected, the tragedy of human deaths was swamped by record-breaking falls on stock markets and forced closures of workplaces interrupting vertical and horizontal supply chains throughout the world.

The fragility of interdependencies and the extreme weakness of contemporary heteronomous systems ignites debates about alternatives: the need to relocalise production; the importance of organising and governing locally to respond more effectively than the state; the need for solidarity as welfare payments and relief are bogged down in red tape and the state selectively funds some people, sectors and activities over others.

By the time this book is published in the European autumn, it is impossible to predict where the world will be. However, we are reasonably certain that analyses and responses offered by the degrowth movement, which we have humbly presented here, will be even more relevant. The degrowth movement offers a radical analysis of a world system that depends on, and is deeply culturally entrenched in an always-more and just-in-time system of production for trade. Most significantly, the degrowth agenda is based in environmental and social justice, seeking to re-politicise societies and open debates on a series of proposals for ways forward.

As Italian intellectual Antonio Gramsci wrote: 'the old world is dying, and the new world struggles to be born: now is the time of monsters'.[2] As this civilisation dies, extremely violent collateral disasters are bringing everybody together. Thin as they might be on the ground in certain places, in other regions transformative movements proliferate. Degrowth spheres of activity and prefigurative formations are under way, decentralised and gradually forming as networks, and alive as works-in-progress attracting greater interest and more members. This cultural climate of chaos and uncertainty has thrown the logic of degrowth into clear relief; suddenly, instead of going against the growth grain, degrowth makes sense.

Moreover, we are satisfied that the degrowth movement conforms to all four strategies that Wallerstein identifies as crucial for this period of transition. First, the degrowth movement sits with 'a process of constant, open debate about the transition and the outcome we hope for'. Here inclusion, social diversity and a multiplicity of voices is crucial. Second, the degrowth movement acknowledges the need for 'short-term defensive action' in order to alleviate the negative effects of the current system and, third, has worked on 'the establishment of interim, middle-range goals that seem to move in the right direction'. Fourth, degrowth advocates and activists are committed to 'a relatively democratic and relatively egalitarian' future. Degrowth principles are amenable to many applications in pragmatic and localised ways, leaving open the form of degrowth and the pursuit of ways to 'discuss it, outline it, [and] experiment with alternative structures to realize it'.[3]

Everywhere, the emergence of movements and experiments to construct a new world can be observed. Our hope is that this book contributes to this upsurge by making the sources and methods of radical system change in the form of degrowth more visible and influential. The pedagogy of catastrophes will play its role: many degrowth predictions or assumptions have been borne out or seem more plausible. The importance of social and environmental justice, of solidarity and localised self-provisioning, have all become clearer. Degrowth solutions have become more relevant and will continue to attract growing interest.

Most of all, we hope that our book initiates and enriches lively debates and practical experiments. More than ever, in these turbulent times, we all need to slow down and think, meet, listen and argue in constructive ways. We need to put our ideas into practice, and critically assess and improve all our practices. That is the invitation made by this book. Debate, and action, must go on!

Vincent Liegey, Budapest, Hungary,
Anitra Nelson, Castlemaine, Australia
16 April 2020

Appendix 1: A Platform For Degrowth

This 'Degrowth: Platform for Convergence' was proposed by attendees of the Constitutional Convention of the Association of Objectors to Growth in Beaugency (France) on 19 September 2009: www.partipourladecroissance.net/?p=6541

Confronted with a multidimensional crisis, we propose a radical, coherent and conscientious proposition: an Objection to Growth.

After the first political campaign Europe-Decroissance (Europe Degrowth) which advocated Degrowth [see definition at the end of this Appendix], do we have the audacity to extend this adventure and to create a political movement, the objectives of which are individual and collective emancipation, well-being and fulfilment achieved through voluntary simplicity?

We consequently do not hesitate to present the outlines of our approach in this endeavour. We invite all like-minded people to join us in developing it collaboratively in order to create a society which is ecologically and socially responsible, humanistic, and democratic.

The 'Growth society' (i.e. business-as-usual paradigm) dramatically creates a combined crisis: an energy crisis, an environmental crisis, a social crisis, an economic crisis, a cultural crisis and a political crisis. Fundamentally, it is the expression of one and the same madness: a world that has become inhuman, as if the belief in 'always more' was enough to give it meaning.

In view of the monumentality of these crises, we offer a new paradigm: an immediate exit from capitalism and productivism through all the practical alternative ways of living life which either exist or are still to be created. These two imperatives constitute our radical approach. We do not want any other form of development, any other type of growth, any other way of consuming or any more productivism. We want to leave behind the religion of Growth which has increased inequality, exhausted resources, wiped out biodiversity and denied human dignity.

Even if infinite growth in a finite world were possible, consumption and productivism would still be absurd.

So how can we create new pathways into different worlds?

In contrast with the classical strategy of taking power as a prerequisite to any change, we propose a radical and coherent idea: The Strategy of the Snail.

Degrowth is not only the aim of such a project, it is also the path and the method.

The political strategy of Degrowth is a strategy of breaking off, in opposition to the strategies of accompaniment.

First, the Strategy of the Snail implies that it is an illusion that acceding to power – whether in a reformative or revolutionary manner – is a prerequisite for changing the world. We do not want to 'seize power', but to act against the dominant structures and ideas by weakening their various powers and to create, without delay, conditions which will enable us to give full meaning to our lives.

Nevertheless, our rejection of the absurd paths of consumerist globalization must not lead us to the dead ends of individualism and isolationism. The Objectors to Growth [see definition at end of the Appendix] (2) intend to be present at all levels of the political spectrum by being active in all the fields/spheres of individual and social emancipation. These include:

- An electoral presence within the traditional political field, avoiding electioneering for it's own sake, but inclusive of demonstrations, petitions, critiques and occasional support for political majorities in order to gain a platform for our ideas and to conduct some social, ecological and political field trials.
- An immediate exit from capitalism effected through experimenting with practical alternatives and alternative counter ideas.
- The project itself, by harmonising utopias into a cohesive whole.

With a view to strengthening our capacity to resist attack, let us re-politicise society and socialize politics.

But since resisting is not enough and something new must be proposed in a Platform, in this platform (which launches our project from its local dimensions into the global) we argue for serene and convivial Degrowth. How?

- Relocalisation constitutes the core of our project; this encompasses housing, travel, production, distribution, exchange and decision-making.
- For the Objectors to Growth, relocalisation is the only sensible pathway for regaining control over our activities, to take care of the land and share common goods.
- The promotion and legalisation of local currencies and other non-speculative exchange systems.
- UAA: An Unconditional Autonomy Allowance (DIA in French: *Dotation Inconditionnelle d'Autonomie*), to put an end to compulsory work and allow for choice of human activity.
- A Maximum Acceptable Income: MAI (in French RMA: *Revenu Maximum Authorisé*), the threshold of which is democratically decided.
- Free public services and usage of socially and ecologically responsible services.
- Free access to responsible use of water, land and every 'essential' resource on the one hand and higher pricing or prohibition of their misuses on the other hand.
- Leaving behind the centrally driven society of over-consumption and energetic [energy] waste; from the nuclear [industry] to the car industry.
- The rejection of the cult of technology which imposes a society of screens and goods instead of a society of personal and direct social ties: those of solidarity, through cooperation and sharing.
- Emancipation of education and cultural models of competition.

- The establishment of democracy by putting an end to media and advertising propaganda, facilitating collective decision-making about economic matters, relocalising the centres of democratic control, setting up protective measures to avoid abuses of power. (i.e. short, revocable and non-cumulative electoral mandates, legislative powers given to the elected representatives, voting rights of a truly universal nature).

To keep on developing this platform ...

(1) Degrowth is either:

- A provocative slogan reminding us that an infinite economic growth in a finite world is not possible.
- A complex set of theories about a new paradigm and thinking about how to make a democratic and peaceful transition to sustainable and desirable societies.
- The name of a new social and political movement (grassroots and also academic and intellectual).

(2) An objector to Growth is a Degrowth proponent.

Appendix 2: The Content of the Unconditional Autonomy Allowance

This is an adapted and updated translation from Vincent Liegey et al., Un Projet de Décroissance, Utopia, 2013, see: www.projet-decroissance.net

The unconditional autonomy allowance offers everyone, from birth to death, enough for a decent and frugal way of life. At the same time the unconditional autonomy allowance is a transition tool towards sustainable and desirable models of societies based on degrowth principles. One of the main challenges is repoliticisation – implementing democratic debates on defining basic needs and how to self-organise locally to satisfy them in sustainable and fair ways.

ASPECTS	AMOUNTS	METHODS AND PROCESSES	TRANSFORMING SYSTEMS AND DECOLONISATION OF IMAGINARY
Right to housing and access to real estate			
Residential accommodation space	A compact and specific maximum allowed space per capita: X m^2	Progressive re-appropriation of access to real estate via requisition laws and participatory deliberation to define local use rights and conditions of use	Energy transition though improvements to insulation; flexibility, multi-functionality and sharing of spaces Challenge prevailing property rights and rehabilitate use rights
Space for social activities			Economic and environmental transition via open relocalisation of production, exchange and other activities
Land surface for agricultural and other productive activities			Agricultural transition to local food provisioning and self-governing territorial autonomy
Rights to basic needs: examples			
Food	Local participatory deliberation over what is 'enough'; based on local estimates, plus deciding what and how to produce and share it	Local currency, local exchange systems (time-bank, reciprocity economies)	Based on direct trade, relocalised and seasonal, less meat-based, organic and sustainable food production Production via agroforestry and agroecology principles
Basic tools such as bikes, furniture, clothes, toys and so on		Temporary use of euro or national currency prior to implementing alternative economic systems	Implementation of open-source, low-tech, hand/home-made, sharing, recycling and makers' workshops Includes all kinds of furniture, clothes, bikes, cargo bikes and trailers, machines, utensils, and so on

CONTENT OF THE UAA ◆ 165

Free access to limited quantities of basic goods			
Water	Depending on local circumstances, enough for a 'meaningful' use Decided by local community	Per capita – X litres or kWh monthly maximum free allocation Easily implemented via meters	Democratic and local re-appropriation of water and energy management Phase out pricing and phase in free access for 'good-usage', and charge for overuse and/or misuse
Energy sources (e.g. fuel, petrol, gas, wood)			Energy transition based on sobriety, effectiveness, and renewable energy (in particular solar thermal) as local, low-tech, and handcrafted as possible toward territorial energy sovereignty
Mobility rights			
Local and short-distance transport	X km per capita allowance Might be phased in gradually depending on transport type	Free local public transport, e.g. limited number of km by train per capita	Rethink urbanisation and dependency on transport as relocalisation of activities is implemented Develop active/soft transport systems (bikes, walk) Right to free public transport gradually reduces following relocalisation to avoid meaningless daily transport
Long-distance transport	X km, per capita allowance (conditional)	X km per capita package (options)	Following open relocalisation, access to long-distance travel, dialogue, cooperation and solidarity all remain open and negotiable Preference for travel by train, bike and sailing boats and visits of long duration

Rights to public services

Health	Free access conditional on revising the content and form of public services (as in column to right)	Preventative approach to medicine through eating and living healthily	As a cultural evolution and transformation, a *progressive implementation* of the unconditional autonomy allowance is necessary, whereby the political imaginary is decolonised. The transition raises continual questions, discussion and debate around the meaning of our life and lifestyles – What do we need to produce? How will these goods and services be produced? In what ways will they be used? How will the hard tasks be shared?
Education		Deschool society (Illich) Skill citizens for autonomous society	
Culture		Arts play a central role in shaping new political imaginary	
Information Communication		Open right to access all kinds of information	
Care and other services		Care of all people's various needs, e.g. for children, people with disabilities and the elderly, funerals, etc.	

Maximum acceptable income as an example of implementing associated measures			
Maximum acceptable income	A multiple (say quadruple) of the minimum income	Taxes on all income over and above the guaranteed minimum income Taxes democratically decided	Associated with radical reform of banking and financial systems: reassessment of public debts with moratorium on debts deemed illegal, unfair or unethical; democratic and transparent governance of central banks and money creation, and other financial and banking systems; implement alternative local/complementary non-speculative currencies linked to transition projects Strict regulation of fiscal evasion, tax havens, and sectors such as military industrial complex, advertising and marketing, mega-infrastructure projects No planned obsolescence

Appendix 3: Implementing the Unconditional Autonomy Allowance: Transitionary Steps

This appendix is adapted and updated from Vincent Liegey et al., Un Projet de Décroissance, Utopia, 2013, see: www.projet-decroissance.net

The unconditional autonomy allowance is not a magical recipe but offers diverse pathways, a coherent convergence of complementary levels and approaches. It is driven by cultural transformation, a decolonisation of our imaginary, rather than being a technical tool to solve all institutional, economic and political problems. It is envisaged as one pathway within an emancipating, democratic and serene transition from growthism to sustainable, desirable, relocalised but connected, open, convivial and autonomous societies. This appendix identifies desirable, realistic steps leaving open questions: What are our basic needs? What, and how, do we satisfy/produce them?

Step	Levers	Achievements and advantages	Barriers and risks	Convergence
0	CULTURAL TRANSFORMATION Adopting a paradigm switch, decolonisation of the growth imaginary precipitates changes in everyday practices – a continuous process	Polls and studies show urgently needed awareness is under way But the growth ideal is still strong, in particular failing to structurally abate growth mania	Change slow due to advertising, lack of time and space for debate in elite-controlled mainstream media, debt slavery (loans) and employment This step needs others to facilitate it, namely democratisation and the acceleration of an emancipating decolonisation of the growth imaginary	All these degrowth steps need to converge with expanding and multiplying exemplary practices and spaces for experimenting with new ways to produce, exchange and make decisions
1	GRASSROOTS PRACTICES Implementation and extension of local concrete alternatives, initiatives, initially pilots and demonstrations, ultimately creating self-governing formations and commoning	Alternative economic structure in formation with useful results leading to developments of appropriate models Visible and scholarly impacts	Activities difficult to integrate with current state and market structures; compete and conflict with mainstream practices Not fully functional until radical transformation to commons governance and commoning economies	Convergence with other social movements necessary for adoption, including in law, for direct governance of commons and for freedom from paid work
2	ALTERING WORK Voluntary reduction of working hours through part-time and sharing roles – allowing free time to support degrowth formations and transition	Addresses current unemployment and precarity Addresses work–life balance Numerous voluntary changes occurring; numbers of people consider this feasible	Need unions and parties to support part-time work and more flexible working conditions Managers of conventional workplaces can be prejudiced against degrowth advocates and activists Unless there is broader change beyond work sphere, elites may end up controlling even more resources	Requires progressive implementation of an acceptable maximum income; democratic re-appropriation of the monetary system (public debt, money creation, fiscal evasion) to re-embed the economy
3	UNCONDITIONAL BASIC INCOME Implementing an unconditional basic income initially depends at least in part on formal monetary flows	Under debate Some successful small-scale pilots Offers opportunities for liberation from work to engage in degrowth transitions	Needs mass, broadspread implementation and measures to avoid abuse, i.e. particular political, cultural and economic supports Challenges in avoiding shocks to food supply-chain given high level of complexity and fragility	
4	PROGRESSIVE DEMONETISATION Unconditional basic income becomes a fully fledged unconditional autonomy allowance	Establishing decentralised and relocalised exchange networks Democratic and peaceful re-embedding of the economy via voluntary democratic deliberation in favour of unconditional autonomy allowance Open relocalisation creates solidarities balancing the logic of subsidiarity and counter-revolutionary power	Might increase inequalities and tensions between territories due to disruptions in trade and financing arrangements	

Notes

PREFACE

All URLs in the Preface accessed 23 March 2020.

1. National Centers for Environmental Information, 'Assessing the global climate in January 2020', News release, 13 February 2020: www.ncei.noaa.gov/news/global-climate-202001
2. Nassos Stylianou and Clara Guibourg, 'Hundreds of temperature records broken over summer', *BBC News*, 9 October 2019: www.bbc.com/news/science-environment-49753680
3. Joanna Partridge, 'Storm Dennis damage could cost insurance companies £225m', *The Guardian*, 20 February 2020: www.theguardian.com/business/2020/feb/20/storm-dennis-damage-could-cost-insurance-companies-225m
4. Tom Griffiths, 'The language of catastrophe', *Griffith Review* 35: www.griffithreview.com/articles/the-language-of-catastrophe/
5. Colin Gourlay, Tim Leslie, Matt Martino and Ben Spraggon, 'How heat and drought turned Australia into a tinderbox', *ABC News*, update 24 February 2020: www.abc.net.au/news/2020-02-19/australia-bushfires-how-heat-and-drought-created-a-tinderbox/11976134
6. Prime Minister of Australia, 'National Royal Commission into Black Summer bushfires established', Media release, 20 February 2020: www.pm.gov.au/media/national-royal-commission-black-summer-bushfires-established
7. Mike Carlowicz, 'Extreme rain douses fires, causes floods in Australia', NASA Earth Observatory, 12 February 2020: https://earthobservatory.nasa.gov/images/146284/extreme-rain-douses-fires-causes-floods-in-australia

1 INTRODUCTION: EXPLORING 'DEGROWTH'

For all chapters, unless otherwise indicated, translations from French to English have been made by Vincent Liegey. All URLs for chapters 1–4 accessed 20 January 2020; URLs for chapter 5 accessed 20 March 2020.

1. P. Ariès, *La Décroissance: Un Nouveau Projet Politique*, Villeurbanne: Editions Golias, 2007.
2. E.F. Schumacher, *Small is Beautiful: Economics as if People Mattered (25 years later ... with commentaries)*, Point Roberts/Vancouver, BC: Hartley & Marks, 1999 [1973].
3. Sally J. Matthews, 'Postdevelopment theory', *Oxford Research Encyclopedia of International Studies*, New York: International Studies Association and Oxford University Press, 2018 (28 August update): https://oxfordre.com/internationalstudies/view/10.1093/acrefore/9780190846626.001.0001/acrefore-9780190846626-e-39
4. 'Ecological footprint', Global Footprint Network: www.footprintnetwork.org/our-work/ecological-footprint/
5. Cited in Graham Maxton, 'We're all economists now ... just don't expect difficult questions', Commentary in *World Economics Association Newsletter* 2(5), October 2012: www.worldeconomicsassociation.org/newsletterarticles/we-are-all-economists-now/
6. Dennis Meadows, Donella Meadows, Jørgen Randers and William W. Behrens III, *The Limits to Growth: A Report for the Club of Rome's Project on the Predicament of Mankind*. New York: Universe Books, 1972.
7. Cited in translation in footnote 10 of Federico Demaria, François Schneider, Filka Sekulova and Joan Martinez-Alier, 'What is degrowth? From an activist slogan to a social movement', *Environmental Values* 22(2), 2013, pp. 191–215, esp. p. 195, referring to M. Bosquet (André Gorz), *Nouvel Observateur*, Paris, 397, 19 June 1972, p. IV. Proceedings from a public debate organised in Paris by the Club du Nouvel Observateur.
8. J. Grinevald and I. Rens, *Demain la Décroissance: Entropie-Écologie-Économie*. Lausanne: Pierre-Marcel Favre, 1979.

9. *Défaire le Développement, Refaire le Monde!* Un numéro spécial exceptionnel, *L'Ecologiste* 6, Winter 2001: www.ecologiste.org/contents/fr/p33.html
10. The programme of the colloquium can be found at: www.lalignedhorizon.net/colloque_unesco.html
11. Serge Latouche, 'A bas le développement durable! Vive la décroissance conviviale!', *S!lence* 280, February 2002. Full text: www.revuesilence.net/epuises/200_299/silence280.pdf
12. Ibid.
13. G. Rist, *The History of Development: From Western Origins to Global Faith*, 3rd edn, London: Zed Books, 2008 [1996]; the quote is from Institut d'études économiques et sociales pour la décroissance soutenable, 'Historique: Histoire du mot décroissance': www.decroissance.org/index.php?chemin=textes/historique
14. Post-growth 2018 Conference, Brussels, Belgium, 18–19 September: www.postgrowth2018.eu/
15. Here we allude to choices between socialism and barbarism that seem to have been raised first by Marxist Karl Kautsky but gained prominence with their use by Rosa Luxemburg. See Ian Angus, 'The origin of Rosa Luxemburg's slogan "socialism or barbarism"', in his Ecosocialist Notebook Comments and Commentary at the Climate & Capitalism website, 22 October 2014: https://climateandcapitalism.com/2014/10/22/origin-rosa-luxemburgs-slogan-socialism-barbarism/
16. Jeroen C.J.M. van den Bergh and Giorgos Kallis, 'Growth, a-growth or degrowth to stay within planetary boundaries?', *Journal of Economic Issues* 46(4), 2012, pp. 909–20.
17. Anon. (1990) *Queers Read This*. Leaflet. New York: Queers, June: www.qrd.org/qrd/misc/text/queers.read.this
18. See Participants in the Economic De-growth for Ecological Sustainability and Social Equity Conference, 'Declaration of the Paris 2008 Conference', 18–19 April: https://degrowth.org/wp-content/uploads/2011/05/Declaration-Degrowth-Paris-2008.pdf
19. Barbara Muraca and Matthias Schmelzer, 'Sustainable degrowth: Historical roots of the search for alternatives to growth in three regions', in Iris Borowy and Matthias Schmelzer (eds) *History of the Future of Economic Growth: Historical*

Roots of Current Debates on Sustainable Degrowth, Abingdon: Routledge, 2017, pp. 174–97, esp. p. 188.

2 DECOLONISING OUR GROWTH IMAGINARIES

1. Herman Daly, 'Growthism: Its ecological, economic and ethical limits', *real-world economics review* 87, 19 March 2019, pp. 9–22: http://www.paecon.net/PAEReview/issue87/Daly87.pdf
2. All quotations in this section are from Nicolas Georgescu-Roegen as cited in Mauro Bonaiuti (ed.), *From Bioeconomics to Degrowth: Georgescu-Roegen's 'New Economics' in Eight Essays*, 1st edn, Abingdon: Routledge, 2014. Data about Georgescu-Roegen is drawn from Jacques Grinevald in Vincent Liegey and Debora Blake, *Décroissance: Hommage à Nicholas Georgescu Roegen*, 2008: https://vimeo.com/14288450
3. Simon Kuznets, 'How to judge quality', *The New Republic*, 20 October 1962.
4. International Energy Agency, 'Data tables', *Global Energy and CO_2 Status Report: The Latest Trends in Energy and Emissions in 2018*, Paris: IEA, 2019: www.iea.org/geco/data/
5. Hugues Ferreboeuf, *Lean ICT – Towards Digital Sobriety*, The Shift Project, March 2019, pp. 4, 8: https://theshiftproject.org/wp-content/uploads/2019/03/Lean-ICT-Report_The-Shift-Project_2019.pdf
6. T. Parrique, J. Barth, F. Briens, C. Kerschner, A. Kraus-Polk, A. Kuokkanen and J.H. Spangenberg, *Decoupling Debunked: Evidence and Arguments against Green Growth as a Sole Strategy for Sustainability*, Brussels: European Environmental Bureau, 2019: https://eeb.org/library/decoupling-debunked/
7. R. Heinberg, *Peak Everything: Waking Up to the Century of Declines*, Gabriola Island: New Society Publishers, 2007.
8. This translation from Ivan Illich, 'Le Genre Vernaculaire' in *Oeuvres Complètes*, vol. II, Paris: Fayard, 2005, appears in Serge Latouche, 'The wisdom of the snail', trans. Ronnie Richards, Slow Food, 22 March 2017: www.slowfood.com/the-wisdom-of-the-snail/
9. Serge Latouche, *L'Autre Afrique: Entre Don et Marché*, Paris: Albin Michel, 1998.

10. Harry S. Truman, 'Inaugural address of Harry S. Truman', The Avalon Project, 20 January 1949: https://avalon.law.yale.edu/20th_century/truman.asp
11. Animata Traore, *Le Viol de l'Imaginaire*, Paris: Fayard/Acte Sud, 2002.
12. Eduardo Galeano, *Open Veins of Latin America: Five Centuries of the Pillage of a Continent*, New York: Monthly Review Press, 1973; Céline Pessi, Sezin Topçu and Christophe Bonneuil, *Une Autre Histoire des 'Trente Glorieuses': Modernisation, Contestations et Pollutions dans la France d'Après-Guerre*, Paris: La Découverte, 2013.
13. For a recent critique, see Jason Hickel, 'The sustainable development index: Measuring the ecological efficiency of human development in the Anthropocene', *Ecological Economics* 167 (January 2020): https://doi.org/10.1016/j.ecolecon.2019.05.011
14. Majid Ranhema, *Quand la Misère Chasse la Pauvreté*, Paris: Fayard/Actes Sud, 2003.
15. Eric De Ruest and Renaud Duterme, *La Dette Cachée de l'Économie*, Paris: Les Liens qui Libèrent, 2014.
16. Ivan Illich, *Tools for Conviviality*, New York: Harper & Row, 1973.
17. Federal Chamber of Automotive Industries, 'September 2019 new vehicle sales figures announced by FCAI', Media release, FCAI website, 3 October 2019: www.fcai.com.au/news/index/view/news/588
18. Serge Latouche, *Jacques Ellul et le Totalitarisme Technicien*, Paris: Le Passager Clandestin, 2013.
19. Serge Latouche, *Bon Pour la Casse: Les Déraisons de l'Obsolescence Programmée*, Paris: Les Liens Libèrent, 2015. The original reads: 'La publicité crée le désir de consommer, le crédit en donne les moyens, l'obsolescence programmée en renouvelle la nécessité.'
20. Hannah Arendt, *Eichmann à Jérusalem*, New York: Viking Press, 1963.
21. Guy Delande, 'Évaluation médico-économique du coût de la fin de vie', *Académie des Sciences et Lettres de Montpellier* 49, 19 March 2018: www.ac-sciences-lettres-montpellier.fr/academie_edition/fichiers_conf/DELANDE-2018.pdf

22. Ivan Illich, *Medical Nemesis. The Expropriation of Health*, London: Calder & Boyars, 1975.
23. André Gorz, *Farewell to the Working Class*, London: Pluto Press, 1994.
24. Éloi Laurent, *Measuring Tomorrow: Accounting for Well-being, Resilience and Sustainability in the Twenty-First Century*, Princeton (NJ): Princeton University Press, 2017, pp. 47–8. David Graeber, *Bullshit Jobs: A Theory*, London: Allen Lane, 2018.
25. Eilis Lawlor, Helen Kersley and Susan Steed, *A Bit Rich: Calculating the Real Value to Society of Different Professions*, London: New Economics Foundation: https://neweconomics.org/2009/12/a-bit-rich
26. See manifesto and details at the student Taking Action for an Ecological Awakening website: https://pour-un-reveil-ecologique.org/en/
27. Thomas Piketty, *Capital in the Twenty-first Century*, Cambridge, MA: Belknap Press of Harvard University Press, 2014.
28. Oxfam, *Fighting Inequality to Beat Poverty: Annual Report 2018–2019*, Oxford: Oxfam, p. 2: https://oi-files-d8-prod.s3.eu-west-2.amazonaws.com/s3fs-public/2019-12/191219_Oxfam_Annual_Report_2018-19.pdf
29. Oxfam, *Extreme Carbon Inequality*, Media briefing, 2 December 2015, p. 1: www.oxfam.org
30. Karl Polanyi, *The Great Transformation*, New York: Farrar & Rinehart, 1944.
31. Serge Latouche, *Les Précurseurs de la Décroissance: Une Anthologie*, Paris: Le Passager Clandestin, 2016.
32. Giacomo D'Alisa, Federico Demaria and Giorgos Kallis, *Degrowth: A Vocabulary for a New Era*, London: Routledge, 2015.

3 DEGROWTH IN PRACTICE

1. 'Le PPLD a été créé comme un parti. Mais depuis plusieurs années déjà, il n'est plus un "parti". Il est dorénavant nommé Parti-e-s Pour La Décroissance (PPLD)', PPLD, 2019: https://www.partipourladecroissance.net/?page_id=4004

2. George Souvlis and Charlie Post, 'Class, race and capital-centric Marxism: An interview with Charlie Post', 19 January 2018, *Salvage*: https://salvage.zone/online-exclusive/class-race-and-capital-centric-marxism-an-interview-with-charlie-post/
3. Kristin Ross, *Communal Luxury: The Political Imaginary of the Paris Commune*, London: Verso, 2015.
4. Marina Sitrin, *Horizontalism: Voices of Popular Power in Argentina*, Oakland, CA: AK Press, 2006.
5. David Graeber, *Bullshit Jobs: A Theory*, New York: Simon & Schuster, 2018.
6. For example, in France, a survey showed that two-thirds of the respondents had moved to more ethical and environmentally friendly consumption habits: Harris Interactive, 'Une étude Harris Interactive pour L'Observatoire Cetelem', Harris Interactive website, 21 February 2018: https://harris-interactive.fr/opinion_polls/theme-1-enquete-23-responsabilite-et-ethique-dans-la-consommation/. Moreover, an Australian Wellbeing Index Survey conducted by Deakin University found that the well-being of retired Australians was 'significantly higher than non-retirees', attributable to post-work time available to them to cultivate and enjoy friendships, shared interests and mutual care: 'Wellbeing Index finds Australians are happiest in retirement', Media release, Deakin University, 6 September 2019: www.deakin.edu.au/about-deakin/media-releases/articles/wellbeing-index-finds-australians-are-happiest-in-retirement
7. Corinna Burkhart, Matthias Schmelzer and Nina Treu (eds) *Degrowth in Movement(s): Exploring Pathways for Transformation*, Hampshire: Zero Books, 2020, p. 11.
8. Elliott W. Martin and Susan A. Shaheen, *Greenhouse Gas Impacts of Car Sharing in North America*, MTI Report 09-11, San Jose, CA: Mineta Transportation Institute, 2010.
9. For examples, see Anitra Nelson, *Small is Necessary: Shared Living on a Shared Planet*, London: Pluto Press, 2018.
10. See ways that the degrowth movement uses such alternative multi-media at the 'Get involved' section of the Degrowth Web Portal: https://www.degrowth.info/en/get-involved/
11. Residents of le Moulin de Rohanne, la Rolandière, les 100 noms, la Hulotte, Saint-Jean du Tertre, les Fosses Noires, la Baraka and Nantes within the Conseil Pour le Maintien des Occupations,

The ZAD Will Survive, ZAD Forever website, p. 6: https://zadforever.files.wordpress.com/2018/03/zad-will-survive-en-pdf.pdf
12. Ibid., p. 11.
13. Ibid., p. 10.
14. For details see 'Biennial international conferences on degrowth for ecological sustainability and social equity', Research and Degrowth (R&D) website: https://degrowth.org/conferences/ and Support Group, 'Reflections on a decade of degrowth international conferences': https://theecologist.org/2018/aug/29/reflections-decade-degrowth-international-conferences
15. See the 2018 EU Post-growth Conference: https://www.postgrowth2018.eu/
16. Eduardo Gudynas in Aaron Karp, 'Contribution to "Roundtable on Vivir Bien"', Great Transition Initiative, 2018: www.greattransition.org/roundtable/vivir-bien-eduardo-gudynas
17. For example, see Degrowth Conference Budapest 2016, 'Principles of the conference': https://budapest.degrowth.org/?page_id=407
18. Giorgos Kallis, Vasilis Kostakis, Steffen Lange, Barbara Muraca, Susan Paulson and Matthias Schmelzer, 'Research on degrowth', *Annual Review of Environment and Resources* 43 (October 2018), pp. 291–316.
19. 'Open letter to François Hollande: Let's transform Aulnay-Sous-bois social tragedy into an ecological and economic transition opportunity for a desirable future', 19 July 2012, PPLD: www.partipourladecroissance.net/?p=7518
20. See Earthworker Cooperative: https://earthworkercooperative.com.au/

4 POLITICAL STRATEGIES FOR DEGROWTH

1. 'The Frog and the Ox', *Fables of Aesop: A Complete Collection*, 27 March 2019 update: https://fablesofaesop.com/the-frog-and-the-ox.html
2. Cornelius Castoriadis, *Philosophy, Politics, Autonomy: Essays in Political Philosophy*, trans. and ed. D.A. Curtis, Oxford:

Oxford University Press, 1991; John Holloway, *Change the World Without Taking Power*, London: Pluto Press, 2010.
3. Adam David Morton, *Unravelling Gramsci: Hegemony and Passive Revolution in the Global Economy*, London: Pluto Press, 2007.
4. For this 1987 speech by Thomas Sankara delivered in French, see AfricaMediaTV, Thomas Sankara – Discours Sur La Dette [Sommet OUA, Addis Abeba] Partie 1/2 (AMTv – AFRIQUE): www.youtube.com/watch?v=FhkqN6KTtJI
5. See Appendix 1, this volume, for 'A platform for degrowth', sourced from the PPLD website: www.partipourladecroissance. net/?p=6541
6. Angelique Chrisafis, 'France braces for *gilets jaunes* anniversary marches', *The Guardian*, 15 November 2019: www.theguardian. com/world/2019/nov/15/france-braces-for-gilets-jaunes-anniversary-marches; K. Willsher, 'Black-clad youths clash with police as *gilets jaunes* mark anniversary', *The Guardian*, 17 November 2019: www.theguardian.com/world/2019/nov/16/paris-police-fire-teargas-on-anniversary-of-gilets-jaunes-protests
7. Hannah Arendt, *Eichmann in Jerusalem: A Report on the Banality of Evil*, London: Penguin Classic, 2006 [1963].
8. Among other works on these topics, these types of models are discussed in various chapters in these collections: A. Nelson and F. Schneider, *Housing for Degrowth: Principles, Models, Challenges and Opportunities*, Abingdon: Routledge, 2018; A. Nelson and F. Edwards (eds), *Food for Degrowth*, Abingdon: Routledge, forthcoming.
9. S. Abuelsamid, 'New vehicles keep getting heavier – or are they?', *Forbes*, 3 January 2019; S.C. Walpole, D. Prieto-Merino, P. Edwards, J. Cleland, G. Stevens and I. Roberts, 'The weight of nations: An estimation of adult human biomass', *BMC Public Health* 12(1), 2012, pp. 439ff; T. Trigg, 'Cities where it's faster to walk than drive', *Scientific American*, 16 May 2015.
10. Patrick Piro, 'Décroissance: Une révolution silencieuse?', 17 February 2016: www.politis.fr/articles/2016/02/decroissance-une-revolution-silencieuse-34137/
11. A quote from *Odoxa*, 'Barometre economique d'octobre français plus ecolos jamais' ['The French are more "green" than ever']: 'Pour préserver l'environnement, les Français sont même devenus des adeptes de la décrois-

sance (54% vs 45%) plutôt que d'une croissance "verte"' ['To preserve the environment, French even become degrowth fans (54% vs 45%) rather than support "green" growth.'] *Odoxa*, 3 October 2019: www.odoxa.fr/sondage/barometre-economique-doctobre-francais-plus-ecolos-jamais/

12. See Philippe Moati, 'L'utopie écologique séduit les Français', 22 November 2019: 'Clairement, l'utopie écologique semble ainsi s'être départie des imaginaires négatifs qui pouvaient être associés à l'idée de décroissance': www.lemonde.fr/idees/article/2019/11/22/philippe-moati-l-utopie-ecologique-seduit-les-francais_6020062_3232.html

13. For details: l'ObSoCo, '3 scénarios pour une utopie – Philippe Moati', Observatoire Société et Consommation, 23 June 2016: http://lobsoco.com/3-scenarios-pour-une-utopie/

14. IMDb, *Tomorrow* (2015); *Demain* (original title)', IMDb: www.imdb.com/title/tt4449576/?ref_=fn_al_tt_1

15. Some key points made in the 2019 IPBES report are that 'most of the globe has now been significantly altered by multiple human drivers, with the great majority of indicators of ecosystems and biodiversity showing rapid decline'; that 'Human actions threaten more species with global extinction now than ever before'; and that 'loss of diversity, including genetic diversity, poses a serious risk to global food security by undermining the resilience of many agricultural systems to threats such as pests, pathogens and climate change'. S. Díaz, J. Settele, E.S. Brondízio, H.T. Ngo, M. Guèze, J. Agard, A. Arneth, P. Balvanera, K. A. Brauman, S.H.M. Butchart, K.M.A. Chan, L.A. Garibaldi, K. Ichii, J. Liu, S.M. Subramanian, G.F. Midgley, P. Miloslavich, Z. Molnár, D. Obura, A. Pfaff, S. Polasky, A. Purvis, J. Razzaque, B. Reyers, R. Roy Chowdhury, Y.J. Shin, I.J. Visseren-Hamakers, K.J. Willis, and C.N. Zayas (eds), *Summary for Policymakers of the Global Assessment Report on Biodiversity and Ecosystem Services of the Intergovernmental Science-Policy Platform on Biodiversity and Ecosystem Services*, Bonn: IPBES Secretariat, 2019, pp. 11–12.

16. John Mecklin, 'Closer than ever: It is 100 seconds to midnight', *Bulletin of the Atomic Scientists*, 23 January 2020: https://thebulletin.org/doomsday-clock/current-time/

5 THE DEGROWTH PROJECT: A WORK IN PROGRESS

1. On distinctions between degrowth and the steady state economy (and room for alliances) see: Joan Martínez-Alier, Unai Pascual, Franck-Dominique Vivien and Edwin Zaccai, 'Sustainable de-growth: Mapping the context, criticisms and future prospects of an emergent paradigm', *Ecological Economics* 69, 2010, pp. 1741–47, esp. pp. 1743–44; I. Cosme, R. Santos and D.W. O'Neill, 'Assessing the degrowth discourse: A review and analysis of academic degrowth policy proposals', *Journal of Cleaner Production* 149, 2017, pp. 321–34, esp. pp. 328–29.
2. V. Liegey, S. Madelaine, C. Ondet and A. Veillot, 'Neither protectionism nor neoliberalism but "open relocalization": The basis for a new International', originally published in French in *Bastamag*, 4 November 2015, 2019, English translation by Dan Golembeski, Un Projet de Décroissance website: www.projet-decroissance.net/?p=2125
3. Ashish Kothari, A. Salleh, A. Escobar, F. Demaria and A. Acosta, *Pluriverse: A Post-Development Dictionary*, Delhi: Authors Up Front and Tulika, 2019, especially for terms such as 'radical ecological democracy', 'ecological *swaraj*', '*buen vivir*' and *sumak kawsay* that emphasise autonomy, a right to basic needs, respecting Earth and a life of dignity.
4. Corinna Burkhart, Matthias Schmelzer and Nina Treu (eds), *Degrowth in Movement(s): Exploring Pathways for Transformation*, Alresford, Hants: Zero Books/John Hunt Publishing, 2020.
5. Nuno Videira, François Schneider, Filka Sekulova and Giorgos Kallis, 'Improving understanding on degrowth pathways: An exploratory study using collaborative causal models', *Futures* 55, 2014, pp. 58–77, esp. p. 59.
6. Vincent Liegey, Stéphane Madelaine, Christophe Ondet and Anne-Isabelle Veillot, *Un Projet de Décroissance: Manifeste pour une Dotation Inconditionelle d'Autonomie* [Degrowth Project, Manifesto for an Unconditional Autonomy Allowance], Paris: Éditions Utopia, 2013, see Un Projet de Décroissance website: www.projet-decroissance.net

7. Jason Hickel, 'Degrowth: A theory of radical abundance', *Real-World Economics Review* 87, March 2019, pp. 54–68.
8. Voting as at 22 February at 'Climate, inequality, hunger: Which global problems would you fix first?', *The Guardian*, 15 January 2020: www.theguardian.com/global-development/ng-interactive/2020/jan/15/environment-inequality-hunger-which-global-problems-would-you-fix-first
9. Matthew Taylor, 'Climate crisis seen as "most important issue" by public, poll shows', *The Guardian*, 19 September 2019: www.theguardian.com/environment/2019/sep/18/climate-crisis-seen-as-most-important-issue-by-public-poll-shows
10. Giorgos Kallis, 'You're wrong Kate. Degrowth is a compelling word (A response to Kate Raworth)', 2 December 2015 post at Oxfam blog 'From Poverty to Power': https://oxfamblogs.org/fp2p/youre-wrong-kate-degrowth-is-a-compelling-word/
11. '2020 Edelman Trust Barometer' (page from which the quotes are taken and where the report can be downloaded), 19 January 2020: www.edelman.com/trustbarometer
12. For more details on and records of this conference, see: www.postgrowth2018.eu/
13. Kothari et al., *Pluriverse: A Post-Development Dictionary*.
14. M. Fisher, *Capitalist Realism: Is There No Alternative?* Zero Books/John Hunt Publishing, 2009, p. 2; Fredric Jameson, *The Seeds of Time*. New York: Columbia University Press, 1994; Slavoj Žižek, 'The spectre of ideology' in *Mapping Ideology*, ed. S. Žižek. New York: Verso, 1994.
15. Hickel, 'Degrowth: A theory of radical abundance'.
16. Diego Andreucci and Salvatore Engel-Di Mauro (2019) 'Capitalism, socialism and the challenge of degrowth: Introduction to the symposium', *Capitalism Nature Socialism*, 30(2), pp. 176–88; also see Burkhart et al., *Degrowth in Movement(s)*, p. 23. An interesting survey, not without certain flaws, is Dennis Eversberg and Matthias Schmelzer, 'The degrowth spectrum: Convergence and divergence within a diverse and conflictual alliance', *Environmental Values* 27, 2018, pp. 245–67.
17. Federico Demara, Giorgos Kallis and Karen Bakker, 'Geographies of degrowth: Nowtopias, resurgences and the decolonization of imaginaries and places', *Environment and*

Planning E: Nature and Space, 2(3), 2019, pp. 431–50, esp. p. 439.
18. Deborah Cowen, *The Deadly Life of Logistics: Mapping Violence in Global Trade*, Minnesota: University of Minnesota Press, 2014.
19. Andrew Jorgenson and Brett Clark, 'Ecologically unequal exchange in comparative perspectives: A brief introduction', *International Journal of Comparative Sociology*, 50(3–4), 2009, pp. 211–14; Prapimphan Chiengkul, 'The degrowth movement: Alternative economic practices and relevance to developing countries', *Alternatives: Global, Local, Political*, 43(2), 2018, pp. 81–95; Beatriz Rodríguez-Labajosa, Ivonne Yánezc, Patrick Bond, Lucie Greyle, Serah Mungutif, Godwin Uyi Ojog and Winfridus Overbeekh, 'Not so natural an alliance? Degrowth and environmental justice movements in the Global South', *Ecological Economics*, 157, 2019, pp. 175–84, esp. p. 180.
20. Kothari et al., *Pluriverse: A Post-Development Dictionary*; A. Kothari, 'Earth Vikalp Sangam: Proposal for a global tapestry of alternatives', *Globalizations* 17, 2020; Demara et al., 'Geographies of degrowth', pp. 440–45.
21. Nelson and Schneider (eds), *Housing for Degrowth*; Nelson and Edwards (eds) *Food for Degrowth*.
22. First North–South Conference on Degrowth-Decrecimiento, 3–7 September 2018, Palacio de Medicina (Former School of Medicine), Universidad Nacional Autónoma de Mexico (UNAM), Mexico City: https://degrowth.descrecimiento.org/
23. Nassim Kahdem, 'Australia has a high rate of casual work and many jobs face automation threats: OECD', Australian Broadcasting Commission, 25 April 2019: www.abc.net.au/news/2019-04-25/australia-sees-increase-in-casual-workers-ai-job-threats/11043772
24. Angela Y. Davis, 'Angela Davis: An interview on the futures of Black radicalism', Verso blog, 11 October 2017: www.versobooks.com/blogs/3421-angela-davis-an-interview-on-the-futures-of-black-radicalism
25. Burkhart et al., *Degrowth in Movement(s)*.
26. Kallis, 'You're wrong Kate'.
27. Martial Foucault, Yann Algan, Daniel Cohen, Elizabeth Beasley and Madeleine Péron, 'Qui sont les Gilets jaunes et leurs soutiens?', Observatoire du Bien-être du CEPREMAP

et CEVIPOF, Sciences Po publications 2019-03, 14 February 2019.
28. Richard Greeman, 'France at the crossroads', *CounterPunch*, 15 January 2020: www.counterpunch.org/2020/01/15/france-at-a-crossroads/
29. Cornelius Castoriadis, *Postscript on Insignificance: Dialogues with Cornelius Castoriadis*, ed. Gabriel Rockhill, trans. John V. Garner. London: Bloomsbury Academic, 2011.
30. A. Guttman, 'Global advertising spending 2010-2019', *Statista*, 8 January 2020: www.statista.com/statistics/236943/global-advertising-spending/
31. Martínez-Alier et al., 'Sustainable de-growth', p. 1744.
32. Karl Polanyi, *The Great Transformation: The Political and Economic Origins of Our Time*, Boston, MA: Beacon Press, 2001 [1944].
33. We must iterate here that the proposals made in this chapter owe much to the work of Liegey et al., *Un Projet de Décroissance*.
34. See especially *Agrarian Justice* in Thomas Paine, *Writings*, ed. Eric Foner, Philadelphia, PA: Library of America, 1993.
35. Louise Haagh, *The Case for Universal Basic Income*, Cambridge: Polity, 2019; Ronald Blaschke, 'Basic income: Unconditional social security for all', in Burkhart et al., *Degrowth in Movement(s)*, pp. 73-86.
36. Matt Orfalea, 'Milton Friedman on guaranteed income/negative income tax/basic income', 26 February 2019: www.youtube.com/watch?v=YLt2X8Zybds
37. Vincent Liegey, Stéphane Madelaine, Christophe Ondet and Anne-Isabelle Veillot, 'A springboard for a Degrowth Project: The citizens' initiative for a basic income', Un Projet de Décroissance website, 25 June 2013: www.projet-decroissance.net/?p=873
38. Vincent Liegey, Stéphane Madelaine, Christophe Ondet, and Anne-Isabelle Veillot, 'A maximum acceptable income: Beyond the symbolic limits' (text commissioned by *Moins! Le journal Romand d'Écologie Politique* [Less! The French-language Swiss Political Ecology Magazine], September 2012), Un Projet de Décroissance website, 7 January 2013: www.projet-decroissance.net/?p=1154

39. Rupert Neate, 'Richest 1% own half the world's wealth, study finds', *The Guardian*, 14 November 2017: www.theguardian.com/inequality/2017/nov/14/worlds-richest-wealth-credit-suisse
40. Hugo Carton, 'Momentum Institute analyses basic income / unconditional autonomy allowance: "Living income for free and egalitarian societies"' (drawn in translation from Hugo Carton, *Le Revenu d'Existence Pour des Sociétés Libres et Égalitaires*, Paris: Institut Momentum, 2013: www.institutmomentum.org/wp-content/uploads/2013/10/Le-revenu-d%e2%80%99existence.pdf), Un Projet de Décroissance website, 10 October 2013: http://www.projet-decroissance.net/?p=1640
41. On commons and degrowth see: Johannes Euler and Leslie Gauditz, 'Commons: Self-organized provisioning as social movements', in Burkhart et al., *Degrowth in Movement(s)*, pp. 128–42; Johannes Euler, 'The commons: A social form that allows for degrowth and sustainability', *Capitalism Nature Socialism*, 30(2), pp. 158–75.
42. Manfred A. Max-Neef, *Human Scale Development: Conception, Application and Further Reflections*, New York, Apex Press, 1991.
43. Claudio Cattaneo, 'How can squatting contribute to degrowth?', in Nelson and Schneider (eds), *Housing for Degrowth*, pp. 44–54.
44. Natasha Verco, 'Christiania: A poster child for degrowth?', in Nelson and Schneider (eds), *Housing for Degrowth*, pp. 99–108.
45. Nadia Johanisova, Tim Crabtree and Eva Fraňková, 'Social enterprises and non-market capitals: A path to degrowth?', *Journal of Cleaner Production*, 38, 2013, pp. 7–16.
46. Jessica Bridger, 'The Kalkbreite co-op complex and Zurich's cooperative renaissance', *Metropolis*, 29 November 2016: www.metropolismag.com/cities/housing/kalkbreite-co-op-zurich-cooperative-renaissance/
47. Samba Mbaye and Marialuz Moreno Badia, 'New data on global debt', 2 January 2019: https://blogs.imf.org/2019/01/02/new-data-on-global-debt/
48. Committee for the Abolition of Illegitimate Debt site: www.cadtm.org/CADTM
49. Timothy M. Lenton, Johan Rockström, Owen Gaffney, Stefan Rahmstorf, Katherine Richardson, Will Steffen and Hans

Joachim Schellnhuber, 'Climate tipping points – Too risky to bet against', *Nature*, 575, 27 November 2019, pp. 592–95.
50. G. Kallis, F. Demaria and G. D'Alisa, 'Degrowth', in James D. Wright (editor-in-chief), *International Encyclopedia of the Social and Behavioral Sciences*, 2nd edn, vol. 6, Oxford: Elsevier, pp. 24–30, esp. pp. 28–29.
51. Manuel Castells (ed.) *Another Economy is Possible: Culture and Economy in a Time of Crisis*, New Jersey: Wiley, 2017; Eduard Nus '15M: Strategies, critique and autonomous spaces', in Burkhart et al., *Degrowth in Movement(s)*, pp. 44–58, esp. p. 53.
52. Andrea*s Exner, Justin Morgan, Franz Nahrada, Anitra Nelson and Christian Siefkes, 'Demonetize: The problem is money', in Burkhart et al., *Degrowth in Movement(s)*, pp. 159–71.

POSTFACE: 'NOW IS THE TIME OF MONSTERS'

1. Immanuel Wallerstein, 'New revolts against the system', *New Left Review* 18, November–December 2020, pp. 29–39, esp. pp. 37–38.
2. This is Gramsci according to Slavoj Žižek, 'A permanent economic emergency', *New Left Review* 64, July–August 2010, pp. 85–95, esp. p 95.
3. Immanuel Wallerstein, 'New revolts against the system', pp. 38–39.

Selected Further Reading and Links

A select list of English-language texts and links to sites on degrowth follows. Our notes contain many more references. You can search for specific topics and authors using the Index to this volume. All links were successfully accessed at 27 February 2020.

Avenel, T. '"A degrowth project": Yesterday's utopia has become today's reality!', *EcoRev' Revue Critique d'Ecologie Politique* 41, 2013, English translation: www.projet-decroissance.net/?p=1230

Bookchin, Murray, *Towards an Ecological Society*, Montreal: Black Rose, 1980.

Carton, Hugo, *Living Income for Free and Egalitarian Societies*, Paris: Institut Momentum, 2013, English translation: www.projet-decroissance.net/?p=1640

Castoriadis, Cornelius, *The Imaginary Institution of Society*, Cambridge: Polity Press, 2005.

Cowen, Deborah, *The Deadly Life of Logistics: Mapping Violence in Global Trade*, Minnesota: University of Minnesota Press, 2014.

D'Alisa, Giacomo, Federico Demaria and Giorgos Kallis, *Degrowth: A Vocabulary for a New Era*, London: Routledge, 2015.

Demaria, Federico, François Schneider, Filka Sekulova and Joan Martinez-Alier, 'What is degrowth? From an activist slogan to a social movement', *Environmental Values* 22(2), 2013, pp. 191–215.

Ellul, Jacques, *The Technological System*, New York: Continuum, 1980.

Graeber, David, *Debt: The First 5,000 Years*. London: Melville House, 2011.

Graeber, David, 'On the phenomenon of bullshit jobs: A work rant', *Strike!* 3, August 2013: https://strikemag.org/bullshit-jobs

Georgescu-Roegen, Nicholas, *The Entropy Law and the Economic Process*, Cambridge, MA: Harvard University Press, 1971.

Gorz, André, *Farewell to the Working Class: An Essay on Post-industrial Socialism*, London: Pluto Press, 1994.

Gorz, André, *Reclaiming Work: Beyond the Wage-based Society*, Cambridge: Polity Press, 1999.

Heinberg, Richard, *The Party's Over: Oil, War and the Fate of Industrial Societies*, Gabriola, BC: New Society Publishers, 2005.

Heinberg, Richard, *The End of Growth: Adapting to Our New Economic Reality*, Gabriola, BC: New Society Publishers, 2011.

Hickel, Jason, *The Divide: A Brief Guide to Global Inequality and its Solutions*, London: Penguin (Windmill Books/Cornerstone Digital), 2018.

Hickel, Jason, 'The sustainable development index: Measuring the ecological efficiency of human development in the Anthropocene', *Ecological Economics* 167 (January 2020): https://doi.org/10.1016/j.ecolecon.2019.05.011

Holloway, John, *Change the World Without Taking Power: The Meaning of Revolution Today*, London: Pluto Press, 2002.

Hopkins, Rob, *The Transition Companion: Making Your Community More Resilient in Uncertain Times*, White River Junction, VT: Chelsea Green Publishing, 2011

Hopkins, Rob and Lionel Astruc, *The Transition Starts Here, Now and Together*, White River Junction, VT: Chelsea Green Publishing, 2018.

Hornborg, Alf, 'How to turn an ocean liner: A proposal for voluntary degrowth by redesigning money for sustainability, justice, and resilience', in Lisa L. Gezon and Susan Paulson (eds) Special Section on Degrowth, Culture and Power, *Journal of Political Ecology* 24, 2017, pp. 425–66.

Illich, Ivan, *Deschooling Society*, New York: Harper & Row, 1970.

Illich, Ivan, *Tools for Conviviality*, New York: Harper & Row, 1973.

Illich, Ivan, *Energy and Equity*, New York, Harper & Row, 1974.

Illich, Ivan, *Medical Nemesis: The Expropriation of Health*, London: Calder & Boyars, 1975.

Jackson, Tim, *Prosperity without Growth: Foundations for the Economy of Tomorrow*, Abingdon: Routledge, 2017.

Kallis, Giorgos, *Degrowth*, Newcastle: Agenda Publishing, 2018.

Kallis, Giorgos, Vasilis Kostakis, Steffen Lange, Barbara Muraca, Susan Paulson and Matthias Schmelzer, 'Research on degrowth', *Annual Review of Environment and Resources* 43 (October 2018): 291–316.

Kempf, Hervé, *How the Rich are Destroying the Earth*, Paris: Green Books, 2008.

Klein, Naomi, *Shock Doctrine: The Rise of Disaster Capitalism*, Picador: New York, 2008.

Lafargue, Paul, *The Right to Be Lazy: Essays by Paul Lafargue*, Chico, CA: AK Press, 2011.

Latouche, Serge, *Farewell to Growth*, Cambridge: Polity Press, 2009.

Liegey, V., S. Madelaine, C. Ondet and A. Veillot, 'Neither protectionism nor neoliberalism but "open relocalization": The basis for a new International', originally published in French in *Bastamag*, 4 November 2015; English translation 2019 by Dan Golembeski, Un Projet de Décroissance website: www.projet-decroissance.net/?p=2125

Martínez-Alier, Joan, Unai Pascual, Franck-Dominique Vivien and Edwin Zaccai, 'Sustainable de-growth: Mapping the context, criticisms and future prospects of an emergent paradigm', *Ecological Economics* 69, 2010, pp. 1741–47, esp. pp. 1743–44.

Nelson, Anitra and François Schneider (eds), *Housing for Degrowth: Principles, Models, Challenges and Opportunities*, London: Routledge, 2018.

Polanyi, Karl, *The Great Transformation: The Political and Economic Origins of Our Time*, New York: Farrar and Reinhart, 1944.

Rist, Gilbert, *The History of Development: From Western Origins to Global Faith*, London: Zed Books, 3rd edn 2008 [1996].

Schumacher, E.F., *Small is Beautiful: Economics as if People Mattered*. London: Abacus, 1973.

LINKS

Primary degrowth sites in English include:

Degrowth – www.degrowth.info/en/

Research and Degrowth (R&D) – https://degrowth.org/

A Degrowth Project, English section – www.projet-decroissance.net/?cat=35

Cargonomia Centre for Research and Experimentation – http://cargonomia.hu/?lang=en

Index

1%, the 140–1
99%, the 128

a-growth 10, 16, 17
academies 74–80
acceleration
　of environmental impacts
　　32–3, 115
　of growth 27–8, 31–2
　of inequalities 45, 115
acceptable maximum income
　140–1, 161, 167tab, 169tab
accumulation 39–41
Adbusters 8, 9, 18–19, 33, 68
advertisements 38, 107–8,
　109fig, 134, 135fig, 162,
　169tab
agency 120–5
airports, protests against 70–1,
　99
algorithms 39, 101
Amazon 133
Ariès, Paul 80, 88
assemblies 55
austerity 3, 5, 147
Australia *xx–xxi*, 37, 84, 93fig,
　128–9
authoritarian temptation 94–6
autonomy *ix,* 89–90, 119, 131

'banality of evil' 39, 106–7
Barcelona, exchange system
　in 150

basic needs and services
　43, 83–4, 108–9, 117,
　120–2, 131, 136–45, 161,
　164–6tab
bike cultures 104–5, 146 *(box)*
bioeconomics theory 24–6
Bookchin, Murray *xi*
Boulding, Kenneth 6
'brands aren't your friends'
　poster 38fig
Bristol Pound 149
bubbles (social) 39, 101–2
Budapest 104, 105, 146 *(box)*
Burkina Faso 94–6

capitalism
　development and 4
　exponential growth in 27–31
　'golden age' of 34
　inequality in 83, 124–5
　money and debt in 147–8,
　　151
　as political construct 46–7
　popular support, lack of for
　　123
　recommendation to exit
　　159–60
　vulnerabilities of 154–5
　see also neoliberalism
car speed example 35
carbon emissions *xvi–xvii*
Cargonomia 146 *(box)*
cars 36–7, 108

Castoriadis, Cornelius *ix,* 47, 119, 133
change-makers 130
Cheynet, Vincent 9, 11
China *xxii,* 145, 147, 154–5
Christiania (Copenhagen) 143
civil disobedience 100
class identity 128–9
Clémentin, Bruno 9, 11
climate change *xix, xx–xxi,* 21, 45, 114
climate justice 68–9, 121–2
collapsology 31
collective deliberation 53–7
collective sphere 62–7, 73
commons and commoning *ix,* 137
Communism 4–5
community currencies 149–50, 161, 164tab
conditionality 147
conferences (on/including degrowth) 74–9
   Barcelona (2010) 74
   Beaugency (2009) 158–62
   Budapest (2016) 77, 129
   EU Parliament (2018) 11–12, 123
   Leipzig (2014) 74, 130
   Lyon (2005) 87
   Manchester (2021) 77, 78
   Mexico City (2018) 128
   Montreal (2012) 74
   Paris (2008) 17–18, 74, 78–9
   UNESCO (2002) 8
   Venice (2012) 74
conservation principle 23
consumption
   over-consumption 5–6, 31–2, 38, 161
   relocalisation and 105–9, 146 *(box)*
conviviality *ix,* 2–3
Cooperativa Integral Catalana 150
cooperatives 55, 65, 143–5
COP 45
Copenhagen, commoning in 143
coronavirus *xxii,* 154–5
counter-productivity threshold *x,* 34–6
cultural hegemony 92–4
Cyclonomia DIY Bicycle Social Cooperative 146

Daly, Herman 21, 24
Davis, Angela 129
debt 95, 138–9, 145–9, 150–1, 167tab, 169tab
decentralised networks 90–2, 116, 137
decline 3, 8, 23
decolonisation of the (growth) imaginary *x,* 12, 34, 47–8, 128, 138, 152, 164–7tab, 169tab
decoupling 30–2, 82
décroissance 2–3, 8–10, 17–18, 154
degrowth
   concept 2–4, 20–1, 48, 116, 162
   of inequality 44–7, 69, 124–33, 140–1
   misinterpretations of 3, 16–17, 49, 80–1
   movement for: internal organisation of 86–92, 137; internationalisation of

74–8; non-violent strategy of 99–101; origin of 20–1, 33; potential pitfalls for 94–9; role and scope of 49–50, 86–7, 93–4; slow progress of 15–17, 52, 89–90
popular support for *xv*, 111–12, 123
in practice: collective sphere 62–7, 73; constraints 58–60, 67–8; individual sphere 57–62, 73; methods (grassroots examples) 51, 60, 63, 64–5, 69–71, 143–545, 146 *(box)*; resistance sphere 60, 67–71, 73
recommendations for: alternative exchange forms 149–52, 164tab; alternative financial system 145–9, 167tab; basic services and needs 141–5, 161, 164–6tab; comprehensive platform 158–67; inclusion 125–33, 156; maximum income 140–1, 161, 167tab, 169tab; open delocalisation *xi–xii*, 105–9, 136–8, 146 *(box)*, 161, 165tab, 169tab; unconditional autonomy allowance 131, 133–40, 142, 151–3, 161, 163, 164–7tab, 169tab; voluntary measures 120–5
terminology 2–3, 7–11, 16–19, 162
thermodynamics and 23–4
see also growth
Degrowth in Movement(s) 63–4

Demain (Tomorrow) 112
democracy 47, 65–6, 82–3, 134, 137, 162
Dennis (storm) *xix*
development 4, 33–4, 159
digital technology 28, 38–9, 161
dignity 142
direct action 53, 100
direct democracy
 in collective sphere 65
 degrowth and 39, 69, 86, 104, 150, 153
 educational system and 47
 exchange systems and 150
 unconditional autonomy allowances and 134, 137, 151–2
 in ZAD 70
disaster capitalism 113–14
Dupuy, Jean-Pierre 114

Earthworker Cooperative 84
eco-collaborative housing 65, 143–5
eco-fascism 94, 111–12
ecofeminism *x*
ecological economics *x*, 24
economic growth *see* degrowth; growth; growthism
Edelman Trust Barometer report 123
Éloi, Laurent 43
emails 36
energy *xvi–xvii*, 23, 27–30, 142, 165tab
entropy law 23, 25
environmental justice 68–9, 121–2
EU Parliament 12, 16–17, 76–7

exchange systems 149–52, 164tab

financial crisis (2007-2008) 5, 45, 113, 147
financial system 145–9, 167tab
fires *xx–xxi*
food 63, 66–7, 75, 164tab
formations, degrowth 66–7, 117, 119–20, 126, 133, 138, 142, 146, 156
fossil fuels 27–30
France
   activism in 69–71, 84, 99, 103–4, 132–3
   degrowth movement in 74, 87–8, 89–90, 123
   health care in 41–2
   media coverage in 109–10
fraternisation 132–3
Friedman, Milton 140
frugal abundance *x–xi*, 2–3, 61–3
frugality 25

GDP (gross domestic product) 26–7, 43–4, 123–4
Georgescu-Roegen, Nicholas 7–8, 22–6
Global South 46–7, 122, 126–8, 131
glocal *xi*
Gorz, André 7, 43, 119
Graeber, David 43–4
Gramsci, Antonio 156
Greece 147
'green growth' myth 30–1
green parties and supporters 80–2, 91
greenhouse gas emissions 28

greenwashing 8, 11, 33
Grinevald, Jacques 8
growth
   acceleration of 27–8, 31–2
   challenging of 4–6
   colonisation of imaginaries by 6, 68
   continued centrality of 82, 123–4
   employment and 43–4
   GDP 26–7, 43–4, 123–4
   incompatibility with averting climate breakdown *xvi–xvii*
   left's support for 81
   natural resource use and 28–32
   as panacea 10
   physical limits to 7, 32–3
   popular support, lack of for 122
   quality vs quantity 26–7
   speed and 98
   see also degrowth
growth society 158–9
growthism 21, 24–5, 48, 86, 127
Guardian 122

happiness and enjoyment 27, 36, 40–1 *(box)*, 59, 61–2, 61fig, 94, 176n6
health care 41–2
heteronomy *ix*
Holloway, John 89–90
Homo economicus 46–7
horizontal organisation 53–7, 90–2
housing 65, 143–5, 164tab
Hungary 104–5, 146 *(box)*

ICT (information and communication technology) 28, 38–9, 161
Illich, Ivan *ix,* 31–2, 34–5, 42, 47, 119, 166tab
imaginaries
  choice between 13fig
  decolonisation of *x,* 12, 34, 47–8, 128, 138, 152, 164–7tab, 169tab
  growth, colonisation of by 6, 68
inclusion 125–33
income
  acceptable maximum 140–1, 161, 167tab, 169tab
  unconditional basic 134, 138–40, 169tab
individual sphere 57–62, 73
inequality
  activism and 125–6
  degrowth and 44–7, 69, 124–33, 140–1
  in emissions production 45
  environmental 68–9, 122
  freedom and degree of 117–18
  of income 35
  maximum income to address 140–1, 161, 167tab, 169tab
  populism and 132–3
  of power 124
  risk of increasing 169tab
  in taxation 148
  see also environmental justice
influence, vs power 94
insignificance 133

Johnson, Boris 114, 132

Jones, Patrick 93fig

Kallis, Giorgos 79, 130
Kantaa 146
Klein, Naomi 113–14
Kuznets, Simon 26

La Ligne d'Horizon 8
labour movement 55
Lafargue, Paul 42–3
L'An 01 (The Year 01) 97
land and housing 65, 143–5, 164tab
Latouche, Serge
  on a-growth 10
  on advertisements 38
  on 'development' 33
  on invention of economy 46
  on neoclassical economics 24
  on re-embedding the economy *xii–xiii*
  in S!lence 9
  on toxic concepts 48
laziness, right to 42–3
Le Monde 112
Le Pen, Marine 114
leadership 56
L'Écologiste (periodical) 8
left, the 55, 81–2
leisure 25
LETS (local exchange trading systems) 149–50
lifestyle menu 144 *(box)*
limits
  physical limits to growth 7, 32–3
  socio-cultural limits 33, 42
Limits to Growth, The (report) 7
local currencies 149–50, 161

low-tech approach *xi*

Macron, Emmanuel 114
Manchester conference (2021) 77, 78
Maris, Bernard 39–40
Marx, Karl 4–5
maximum income 140–1, 161, 167tab, 169tab
Moati, Serge 112
municipalism *xi*

natural disasters 114
neoliberalism 46, 83, 113–14, 132–3
networks 66–7, 90–2, 116
New Economics Foundation (NEF) 44
non-hierarchical organising 53–7, 90–2, 116
non-violent activism 53, 99–101, 119

Obsoco (Observatoire Société et Consommation) 111–12
occupations 70–1
Occupy movement 15–16, 50, 52, 55
Odoxa 111
open relocalisation *xi–xii*, 105–9, 136–8, 146 *(box)*, 161, 165tab, 169tab
organising (by movements) 53–7, 90–2, 116
over-consumption 5–6, 31–2, 38, 161
Oxfam 45, 123

Paine, Thomas 139
'peak' oil 28–30

Pearson, Karl 22
Piketty, Thomas 45, 113
planned obsolescence *xii*, 38
Platform for Convergence 97–8
pluriverse *xii*, 117
Polanyi, Karl *xii–xiii*, 46, 152
Politis 109–10
populism 101, 132
Post, Charlie 53–4
post-development *xii*
postgrowth 18
power
  autonomy vs heteronomy *ix*
  degrowth movement and 88–92, 94, 97, 159–60
  local/municipalism *xi*
  unconditional autonomy allowance and 153
  see also autonomy; democracy; direct democracy
PPLD (Le Parti Pour La Décroissance) 52
praxis 84–5
precarious work and workers 43, 118, 128–9, 169tab

Rabhi, Pierre 95
Rahnema, Majid 34
rationing 123
re-embedding of the economy *xii–xiii*, 152, 169tab
rebound effect 36–7
reflection, sphere of 72–4
Regulation School 134
regulations and laws 107–8, 123, 134, 164tab
relocalisation *see* open relocalisation
renewable energy 29–30
Rens, Ivo 8

repoliticisation 132–3, 163
research 78–80
resistance, sphere of 60, 67–71, 73
revolutionary socialism 54
Rist, Gilbert 10
rural activists 51

salaries, vs value of job 44
Salvage (journal) 53–4
Sankara, Thomas 94–6
Schumpeter, Joseph 21
scientists *xvii*
Seeger, Pete 108, 109fig
shock doctrine 113–14
simplicity 25, 58–9, 61–2
S!lence (magazine) 8, 9
smartphones 37–9
snail example 31–2, 34–5
snail, strategy of the 97–9, 159–62
social media 101–1
Special Patrol Group (UK) 134, 135fig
speed 34–5, 96–9
state, the
  activists and 92, 131, 137–8, 143
  displacement of 92, 137–8
  growth promotion by 123–4
  potential role of 134, 143
steady state economy 24, 117
structural adjustments 147
students 129–30
subsidiarity 137, 169tab
surveys 101, 111–12, 122–3, 176n6
sustainability 82
sustainable development 2, 8–9, 24, 30, 33, 123–4

sustainable lifestyle menu 144 *(box)*
SUVs 37, 108

tax evasion 103–4, 113, 148, 167tab
tax, negative 140
taxes, protests against 103–4, 133
technician totalitarianism 37–8
technological fixes *xvi–xvii, xxii–xxiii*
thermodynamics 23–4
time-banking 149
top-down approaches 94–6
Totnes Pound 149
tourist and fisherman story 40–1 *(box)*
transhumanism *xiii*
transparency 106–8, 132, 151
transportation 68, 165tab
trickle-down economics 45
Truman, Harry 33
Trump, Donald 114, 132

UN Climate Change Conference (COP 21) 45
unconditional autonomy allowance 131, 133–40, 142, 151–3, 161, 163, 164–7tab, 169tab
unconditional basic income 134, 138–40, 169tab
unemployment 43–4, 81, 82, 84, 169tab
UNESCO 8
unions 6–7, 81, 133, 169tab
Upper Volta 94–6
urban activists 51
utopian thinking 94

utopias 62, 70–1, 94, 111–12, 116, 160

violence 99–101

Wallerstein, Immanuel 154
water supply 141–2, 165tab
wealth 24, 45
wildfires *xx–xxi*
women's liberation movement 55
work, role of 42–4, 138–9, 169tab

working groups 55–6
working hours 6, 59, 169tab, 176n6
Wright, Erik Olin 4

yellow vest movement 84, 103–4, 132–3

ZAD (Zone à Défendre) 69–71, 99
Zsamboki Biokert 146
Zurich, cooperative housing in 143

Thanks to our Patreon Subscribers:

Abdul Alkalimat
Andrew Perry

Who have shown their generosity and comradeship in difficult times.

Check out the other perks you get by subscribing to our Patreon – visit patreon.com/plutopress.
Subscriptions start from £3 a month.

The Pluto Press Newsletter

Hello friend of Pluto!

Want to stay on top of the best radical books we publish?

Then sign up to be the first to hear about our new books, as well as special events, podcasts and videos.

You'll also get 50% off your first order with us when you sign up.

Come and join us!

Go to bit.ly/PlutoNewsletter

Printed in Great Britain
by Amazon